Springer Texts in Business and Economics

Springer Texts in Business and Economics (STBE) delivers high-quality instructional content for undergraduates and graduates in all areas of Business/Management Science and Economics. The series is comprised of self-contained books with a broad and comprehensive coverage that are suitable for class as well as for individual self-study. All texts are authored by established experts in their fields and offer a solid methodological background, often accompanied by problems and exercises.

Dietmar Ernst • Joachim Häcker

Derivatives

Options and Futures

Springer

Dietmar Ernst
University of Applied Sciences
Nürtingen-Geislingen (HfWU)
Nürtingen, Germany

Joachim Häcker
Munich University of Applied Sciences
Munich, Germany

ISSN 2192-4333 ISSN 2192-4341 (electronic)
Springer Texts in Business and Economics
ISBN 978-3-031-85821-5 ISBN 978-3-031-85822-2 (eBook)
https://doi.org/10.1007/978-3-031-85822-2

This book is the translation of German textbook 'Derivate' originally published by UTB.

This Springer imprint is published by the registered company Springer Nature Switzerland AG
The registered company address is: Gewerbestrasse 11, 6330 Cham, Switzerland

Philosophy of the Book

In this book, we analyze derivatives from the perspective of financial modeling. In financial modeling, the central questions from the financial industry are mapped and solved holistically with the help of software. We use Excel as our software. Excel is very variable, very widely used, the basic functions are relatively easy to learn, and there are good control options and relatively large amounts of data that can be processed. It must be said, however, that Excel cannot replace a database when it comes to large data sets.

To apply data analytics to big data as a whole, the use of higher-level programming languages, such as Python, is recommended. However, we want to make our book accessible to a wide range of readers, which is why we decided on Excel and not Python. As mentioned in the title, the reader should be able to understand how options and futures work step by step. The book's clear focus is on options due to the holistic design of the book. Step by step, a control cockpit is built from pricing to strategies. Our approach of financial modeling based financial engineering is in the foreground. To present this entire process not only for options but also for futures would go beyond the scope of the book. Therefore, we have limited ourselves to option strategies.

You can download the Excel spreadsheets discussed in this book from the publisher's website. These spreadsheets are based on the teaching concept of Financial Modeling based Financial Engineering, which consists of the following four steps:

Step 1: The assignment is defined. This is where the exact task is outlined.
Step 2: Background information on this topic is listed here.
Step 3: Here you will get to know the respective formula with which the result of the respective assignment can be calculated.
Step 4: The respective implementation in Excel is shown there. At the end of each assignment, you will see the Excel result with all figures in an Excel screenshot.

The framework of the Excel file is analogous to the Excel files listed in the book in terms of structure and appearance. You can therefore first work through the contents of step 1 and step 2 yourself by reading the relevant sections in the book. We then recommend that you transfer the formulas from the book into the downloaded Excel files according to step 3.

In step 4, you then complete the cells belonging to the respective assignment in the corresponding folder. Finally, you can compare your result with the Excel screenshot in the book. If you notice any discrepancies, you can go back to step 2 and keep making changes until your Excel folder looks exactly like the screenshot in the book.

Contents

List of Abbreviations

$\overline{\mu}$	Mean value
C	Call price
D	Dividend yield
d	Sink factor
d_M	Monthly reduction factor
e	Euler's number
i	Respective period
ITM	In the money
n	Number of periods, remaining term
N(d)	Cumulative standard normal distribution
$N(d_1)$	Standard normal distribution (d_1)
$N(d_2)$	Standard normal distribution (d_2)
OTM	Out of the money
P	Price of the put
r_f	Risk-free interest rate
r_i	Individual returns
S	Spot price, Underlying
s/σ	Standard deviation
s2/(σ2)	Variance
Strike	Strike price exercise price
T	Period
U	Slope factor
u_M	Monthly slope factor
V	Implied volatility
X	Exercise price (strike price) of the option
y	Convenience yield
Z	Coupon yields for futures

Part I
Fundamentals and Pricing of Options and Futures

Part I covers the following topics:

- Chapter 1: Fundamentals of Options
- Chapter 2: The Black-Scholes Model incl. the Greeks
- Chapter 3: Fundamentals and Pricing of Futures

1.1 Initial Situation

You are an investor and would like to invest in Pharma Group. The following data for the options you have chosen on the Pharma Group share are given:

Exercise price call	EUR 68
Exercise price put	EUR 72
Price (spot price)	EUR 70
Dividend yield	4.85%
Implied volatility	30%
Term of the option in years	1
Risk free interest	1%

Chapter 1
Basics of Options

1.1 Assignment 1: How Options Work

Task
Explain what is meant by the term option and differentiate a stock option from a stock.

Content
An option is understood to be the securitized right, but not the obligation, to buy (call option) or sell (put option) a certain number (contract size) of an object (underlying) admitted to options trading within a certain period (option period or term).

The seller of an option grants the buyer the right to buy or sell an instrument on a specified date. This option right is granted by the seller of an option to the buyer against payment of a monetary amount (option premium).

Since, from the perspective of the option holder, there is no obligation to exercise the option, the opportunities or risks resulting from the option are distributed asymmetrically. This means that the profit opportunities are theoretically open to the upside (upside potential), while the maximum loss can be limited to the option price payable at the beginning of the option term (downside risk). Due to this fact, one also speaks of an asymmetric risk distribution. Since when have options existed?

The first forms of futures markets emerged in India as early as 2000 BC. Records from the time of the Roman Empire and the Phoenicians also provide evidence of the use of futures. Thales of Miletus was already working with options on olives or on the olive presses around 500 BC. In the Middle Ages, commodity futures transactions were particularly common in England and France. The goods mostly came from Asia and were delivered months later. Around 1630, there was a lively options

Supplementary Information The online version contains supplementary material available at https://doi.org/10.1007/978-3-031-85822-2_1.

D. Ernst, J. Häcker, *Derivatives*, Springer Texts in Business and Economics,
https://doi.org/10.1007/978-3-031-85822-2_1

trade on tulip bulbs in the Netherlands. For example, a buyer of the tulip bulb "viceroy" paid 2 carts of wheat, 4 carts of rye, 4 fat oxen, 8 fat pigs, 12 fat sheep, 2 barrels of wine, 4 tons of beer, 1000 pounds of cheese, a silver cup, a bed, and a suit of clothes. Trading was not done through the Amsterdam Stock Exchange but in a great many speakeasies. In 1848, the "Chicago Board of Trade (CBOT)" was founded. In this year, for the first time in history, a standardized futures contract was traded. In 1973, the fixed exchange rates of currencies were abolished (end of Bretton Woods). As a result, volatilities increased. In this context, the volatility of interest rates also increased. The first financial futures contract, an interest rate future, was introduced in Chicago to hedge interest rates. This was called the birth of financial futures. While the first currency futures on the then seven major currencies were traded at the Chicago Mercantile Exchange (CME) in 1972, the first contract on the S&P 500 took place at the CME in 1982. The German Futures Exchange was founded in 1988. It merged with the Swiss SOFFEX in 1992 to form EUREX. EUREX is still one of the largest derivatives exchanges in the world.

What are the main differences in options?

1. The buyer of an option has a long position, and the seller has a short position (= writer).
2. European options can only be exercised at the end of the term, while American options can be exercised during the entire term.
3. With regard to settlement, a distinction can be made between cash settlement and physical delivery (delivery or takeover of the underlying asset, also known as "physical delivery").
4. Options can be based on various underlying assets.
5. Options can be standardized or customized.

Ad 1: The buyer of an option has a long position, while the seller has a short position or is called a writer. The buyer (long) always has an option writer (short). The writer has no right of choice and is bound to the will of the buyer. In return, he receives the premium payment from the buyer. He has thus entered into the obligation (he has no right of choice) to deliver (in the case of a call option) or to take over (in the case of a put option) the specified quantity of the underlying asset at the agreed time and price in the event of exercise.

Ad 2: Options that can be exercised during the entire term are called American-style options. As a rule, these are options denominated in single stocks. Options that can only be exercised at the end of their term (last trading day) are known as European-style options. This type of option is mainly found in index options. In general, most options are American-style options. European options play a minor role in practice outside of index options. However, due to their less abstract mathematical construction, we will use European options for financial modeling-based pricing of options in the following.

Ad 3: A cash settlement always takes place if no physical settlement is possible or desired. This is the case, among others, for index options. In this case, the difference between the underlying and the strike price is settled in cash. The physical settlement of an underlying occurs, for example, in the case of a stock option.

Ad 4: Options may also differ with regard to the underlying asset. This indicates which goods or which security the option is based on. The technical term underlying is also used. The underlying is thus the trading object on which a forward transaction is based. The underlying instruments of options are diverse, such as shares, bonds, commodities, indices, interest rates, currencies, and other forward transactions.

Ad 5: Standardized option contracts may be traded on a futures exchange. Options, on the other hand, which are individually structured, are concluded and traded only between the contracting parties without the intervention of a futures exchange. They are also known as over-the-counter options (OTC options for short).

Why does an investor buy a call option on Pharma Group, for example?
Based on his analysis, the buyer assumes that the share price will rise. Instead of making a direct investment in the Pharma Group share, he buys call options. Due to the lower capital investment, he can buy significantly more options than shares. If the price movement he expects occurs, he profits significantly more from the leverage effect than if he had bought the share directly.

Why does an investor buy a put option on the Pharma Group, for example?
Analogous to the above case, the investor may assume falling prices based on his analysis. The more the Pharma share price falls, the more profit the investor can make. The profit is only capped if the Pharma Group becomes insolvent, and thus the share price is zero. While the buyer of a call option could still alternatively buy the Pharma Group share, the buyer of a put option has an advantage over a seller of a Pharma Group share (who only limits the loss) in that he can make a profit even if the share price falls.

Furthermore, numerous other cases are conceivable (more on this in Part II). For example, an investor holds Pharma Group shares. He fears that the share price could fall. To hedge his portfolio, he buys put options on Pharma Group shares. This enables him to compensate for his losses if the Pharma Group share price falls. This is because the performance of a put option moves in the opposite direction to the underlying Pharma Group share.

In the event that the price of the Pharma Group share rises again as expected, the purchased puts lose value. The investor consequently realizes a loss, which is, however, limited only to the option position. However, if the share price rises more than the loss, the investor is compensated for this loss and can profit from further price gains despite the downside hedge. If he had sold the shares, he would no longer have been able to profit from the increase.

An option can be accrued against a share as follows (Fig. 1.1):

1.2 Assignment 2: Financial Modeling-Based Financial Engineering

Task
Apply the top 10 financial modeling standards when building models.

Purchase of a(n)...	... Share	... Option on the Share
Relevance	Partial ownership in a company.	The right to buy a share (call) or to sell a share (put) at a predetermined price.
Chances	Participation in share price increase and dividends.	Profit in the case of falling share prices (putt); Profit in the case of rising share prices (call).
Risks	Dependence on the economic success of a company. Complete loss is rather rare.	Increased risk, complete loss possible, additional credit risk of issuer.
Additional rights	In some sense, a shareholder becomes an „entrepreneur". Depending on the type of equity, his rights can vary slightly.	The buyer of an option will always remain an investor. He has no additional rights.
Term	Unlimited.	Limited.

Fig. 1.1 Comparing shares and options. *Source: Ernst and Häcker (2017): page 854*

The Top-10 Financial Modeling Standards		
1.	Define the modeling purpose	✓
2.	Separate the proplem into independent subunits (modules)	✓
3.	Provide a graph of the flow of data and the model structure	✓
4.	Separate inputs from outputs	✓
5.	Choose an unified layout for the worksheet	✓
6.	Use unified formatting	✓
7.	Avoid complex formulas and use only one type of formula	✓
8.	Avoid circular references	✓
9.	Work with control functions	✓
10.	Present the results professionally	✓

Fig. 1.2 The top 10 financial modeling standards. *Source: Ernst and Häcker (2017): page 24*

Content

The approach shown here can be called "financial modeling-based financial engineering." What is meant by this? When creating the structure of the model, compliance with the top 10 financial modeling standards in particular is of central importance (Fig. 1.2):

The following is a brief example of how the top 10 financial modeling standards are applied in this Excel workbook.

Fig. 1.3 The module structure of the Excel workbook

Ad 1: The aim of the book is to build a model that can be used to choose the right option strategy from an investment or hedging point of view, depending on the market situation.

Ad 2: The basic question of Part I (What are options and futures, and how can options and futures be priced?) is divided into the following 13 modules (Here: modules = Excel worksheets) (Fig. 1.3).

Ad 3: This is not shown here for reasons of simplification. How such a sketch can look is shown in the video below at 2:25.

Ad 4: Exactly one input worksheet is created. All other worksheets are output worksheets. Some output cells can—if reasonable—be listed in the input worksheet. However, an input-to-output worksheet is the dead sin of financial modeling-based financial engineering.

Ad 5: In Fig. 1.4, for example, it is clear that the relevant figures are listed in column E in each case.

Ad 6: Each subject block in Fig. 1.4 is separated by a black box, for example.

Ad 7: The option price tree in worksheet 7 can be created quickly by simply dragging the upper-right formula down, and the cell in the second column from the right is also simply dragged down and then to the left.

Ad 8: The output worksheets are linked to the input worksheet but not by circular references to each other.

Ad 9: In Worksheet 10, cell C25, check that the left side corresponds to the right side of the put-call parity.

Ad 10: With the folder "Summary Option" (see Fig. 1.5), the central results of this chapter and Chap. 2 can be presented.

Display in Excel

In Fig. 1.4, it is clear that, apart from two cells, there are only inputs in the input worksheet. All these cells have been marked with the color orange. Furthermore, there are two links highlighted in blue color:

E38 = E30
E52 = E5

In the following output worksheets, there are no inputs (orange) but links to the input worksheet (blue) and cells with formulas (green). Each number has a color!

1.3 Assignment 3: Value Drivers of Options

Task

Name and discuss the value drivers of the options on Pharma Group?

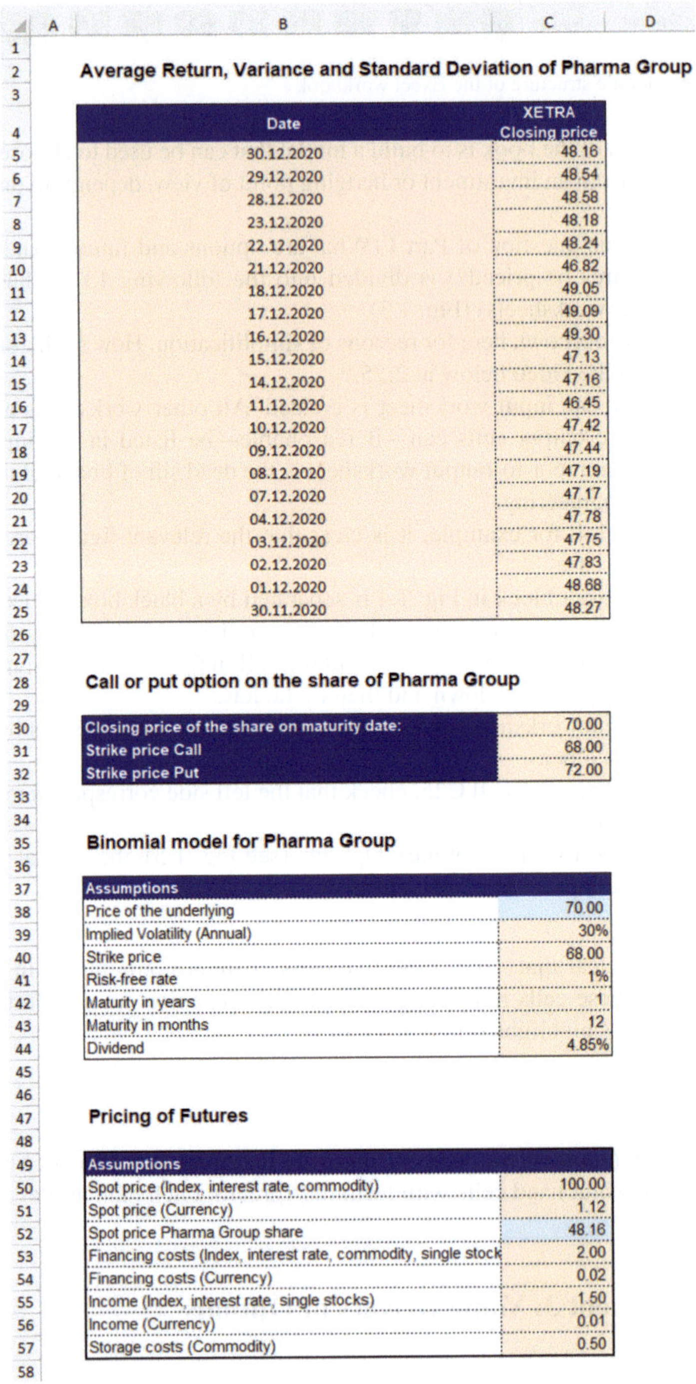

Average Return, Variance and Standard Deviation of Pharma Group

Date	XETRA Closing price
30.12.2020	48.16
29.12.2020	48.54
28.12.2020	48.58
23.12.2020	48.18
22.12.2020	48.24
21.12.2020	46.82
18.12.2020	49.05
17.12.2020	49.09
16.12.2020	49.30
15.12.2020	47.13
14.12.2020	47.16
11.12.2020	46.45
10.12.2020	47.42
09.12.2020	47.44
08.12.2020	47.19
07.12.2020	47.17
04.12.2020	47.78
03.12.2020	47.75
02.12.2020	47.83
01.12.2020	48.68
30.11.2020	48.27

Call or put option on the share of Pharma Group

Closing price of the share on maturity date:	70.00
Strike price Call	68.00
Strike price Put	72.00

Binomial model for Pharma Group

Assumptions	
Price of the underlying	70.00
Implied Volatility (Annual)	30%
Strike price	68.00
Risk-free rate	1%
Maturity in years	1
Maturity in months	12
Dividend	4.85%

Pricing of Futures

Assumptions	
Spot price (Index, interest rate, commodity)	100.00
Spot price (Currency)	1.12
Spot price Pharma Group share	48.16
Financing costs (Index, interest rate, commodity, single stock	2.00
Financing costs (Currency)	0.02
Income (Index, interest rate, single stocks)	1.50
Income (Currency)	0.01
Storage costs (Commodity)	0.50

Fig. 1.4 Creation of the input worksheet

Call option of Pharma Group						Put option of Pharma Group	
Without dividends	in €	With dividends				With dividends	
		European	in €	American	in €	American	
Binomial Model (1 step)	11.59	Binomial Model (6 steps)	7.50	Binomial Model (12 steps)	8.11	Binomial Model (12 steps)	8.41
Black Scholes Model	9.59	Binomial Model (12 steps)	7.60	Market price according to EUREX	7.78	Market price according to EUREX	8.43
		Black Scholes Merton Model	7.67				

Fig. 1.5 Summary of results from this chapter and Chap. 2 (folder: "Summary option")

Content

In the valuation of Pharma Group options, there are the following five value drivers:

1. The *price of the underlying*: This is currently EUR 70.
2. *Volatility*: This amounts to 30%. As a premise, it is assumed that this remains constant during the term. This is the so-called implied volatility and not the historical volatility. It is derived by iteration from the traded options.
3. The *risk-free interest rate*: This is assumed here to be 1%, which corresponds to the money market interest rate for a AAA credit rating.
4. The *dividend yield*: This corresponds to 4.85%.
5. The *remaining term of* the call or put option is 1 year.

Ad 1: Price of the Underlying

The price of the underlying (in the case of an exchange-traded underlying such as the Pharma Group, this corresponds to the share price, i.e., the price of the underlying) has the greatest influence on the option price, as the option moves simultaneously with it. The price of a call therefore increases when the price of the underlying increases. Conversely, it falls if the underlying asset becomes cheaper. In the case of a put, this relationship is exactly reversed: the put becomes more expensive when the underlying asset falls and cheaper when it rises.

This also explains why a derivative (as a derivative of an underlying structure) is directly dependent on the price of the underlying structure. Consequently, a change in the price of the underlying is also associated with a change in the price of the derivative (see Chap. 2: Greeks—Delta).

Ad 2: The Volatility

Volatility σ (Latin volare = to fly) is a statistical measure of the intensity of fluctuations of an underlying asset within a certain period (aggregate total risk) around its mean value.

Volatility only indicates the extent of the fluctuations, not their direction. With a volatility of 10 and a mean value of 100, the underlying fluctuates between 90 and 110 with a probability of 68.26%. Volatility here corresponds to the standard deviation (s or σ), which is defined as the square root of the average mean square deviations of an underlying value from the mean. The standard deviation measures how much the individual returns of the periods fluctuate around the mean. The squared standard deviation, s2 or σ2, is also referred to as the variance. In principle, it can be stated that the higher the volatility, the higher the price of an option. This principle is derived from the reasoning behind this is that high volatility also a high degree of risk.

Basically, the following two types of volatilities can be distinguished:

1. Historical volatility
2. Implied volatility

Ad 1: Historical volatility is a measure of past volatility data of an underlying asset. The historical volatility can refer to different time intervals, such as days, weeks, months, or years.

Ad 2: Implied volatility exists if the volatility represented in the option is a volatility. Implied volatility is the expected future fluctuation of the underlying around its mean value. It is also referred to as perceived volatility, as it is present in the market. The implied volatility reflects the current market opinion, which, depending on the term of the option or the choice of strike price, can vary greatly from the historical volatility may deviate. To find out the implied volatility, it must be derived from the currently traded options. This is done using the Black-Scholes formula and the iteration method (see Chap. 2).

A simple way to get an overview of the current implicit volatility of options, for example, is provided by the volatility index VDAX-NEW of Deutsche Börse Group. The VDAX-NEW measures the implied volatility for the DAX. It indicates, in percentage points, the volatility to be expected for DAX in the next 30 days (see: https://www.boerse-frankfurt.de/wissen/lexikon/vdax-new).

Ad 3: The Risk-Free Interest Rate

The risk-free interest rate is understood here as the money market interest rate. In principle, it corresponds to an AAA credit rating, i.e., an interest rate with no default risk.

When selecting an appropriate risk-free interest rate for an option in practice, other factors, such as the term of the option and its currency, are to be considered.

When the risk-free interest rate rises, the price of a call and the price of the put decreases. What is the reason for this? On the one hand, it is with the market interest rate disadvantages, on the other hand, with the basic circumstances of the option types themselves. The price differences between the direct investment and the investment in the option are balanced.

Let us illustrate this with the example of a call option: The direct investment in an asset would require that we use an investment amount of X today. This capital would then be tied up in the asset and would not be available for free disposal. However, if instead of the asset we buy an option on the asset, which involves a significantly lower capital commitment (e.g., 10% of X), we have a liquidity advantage and can invest the excess liquidity elsewhere.

This investment brings us additional income. However, since arithmetically the investment in an option must be equivalent to the investment in the asset itself (assuming an efficient capital market), the difference is made up via the market price in the option. Therefore, a rising risk-free interest rate makes the price of a call more expensive.

Ad 4: The Dividend Yield

Dividend payments have a direct impact on the price of the underlying. Due to their direct influence, they lead to falling prices for calls and rising prices for puts. The reason for this lies in the primary effect of distributions or dividend changes on the share price: due to the dividend distribution, the share price falls, causing the call price to fall and the put price to rise.

Dividends are mistakenly viewed by some market participants as a form of "additional income." However, dividends represent only a partial repayment of the capital employed. This is also illustrated by the fact that the share price is reduced by the dividend yield on the day after the dividend is paid. If the company did not pay a dividend, the market capitalization of the company would not be reduced by the dividend either.

Ad 5: The Remaining Term

The shorter the term of an option, the lower its price. The chance that the option will end in the money becomes smaller, and the probability of a worthless expiration increases. The remaining term is, therefore, a serious influencing parameter for options that are out of the money or directly at the money. Options that are deep in the money do not have a high sensitivity to the remaining time to expiration because the time value function has already been fulfilled. Thus, deep in-the-money options only have a comparatively lower time value premium (agio).

Important Formulas

Historical volatility is generally indicated by the standard deviation. The formula for the standard deviation (related to the population) is the square root of the average mean squared deviations of an underlying value from the mean and is:

$$S = \sqrt{\frac{1}{n}\sum_{i=1}^{n}\left(r_i - \bar{\mu}\right)^2}$$

The standard deviation s corresponds to the square root of the variance and shows how much the individual returns (r_i) of the periods (n) fluctuate around the mean value ($\bar{\mu}$). The return in the respective period i is calculated by dividing the natural logarithm of the quotient of the share price in period T (P_T) by the share price in period 0 (P_0).

$$r^S = \ln\left(\frac{P_T}{P_0}\right)$$

Here, the continuous rate of return (r^s) is calculated rather than the discrete rate of return, since the normal distribution premise underlies it.

Display in Excel

In column C, the XETRA closing prices of the Pharma Group share for the month of December 2020 (including 11/30/2020) were downloaded. In column D, the steady return is calculated. The cell D5 results as =LN(C5/C6). After that, this cell

Average Return, Variance and Standard Deviation of Pharma Group

Date	XETRA Closing price	Continuous rate of return
30.12.2020	48.16	-0.8%
29.12.2020	48.54	-0.1%
28.12.2020	48.58	0.8%
23.12.2020	48.18	-0.1%
22.12.2020	48.24	3.0%
21.12.2020	46.82	-4.7%
18.12.2020	49.05	-0.1%
17.12.2020	49.09	-0.4%
16.12.2020	49.30	4.5%
15.12.2020	47.13	-0.1%
14.12.2020	47.16	1.5%
11.12.2020	46.45	-2.1%
10.12.2020	47.42	0.0%
09.12.2020	47.44	0.5%
08.12.2020	47.19	0.0%
07.12.2020	47.17	-1.3%
04.12.2020	47.78	0.1%
03.12.2020	47.75	-0.2%
02.12.2020	47.83	-1.8%
01.12.2020	48.68	0.8%
30.11.2020	48.27	

Result	
Average Return	-0.0114%
Standard Deviation	1.7988%
Variance	0.0324%

Fig. 1.6 Average return, standard deviation, and variance of Pharma Group closing prices in December 2020

can be drawn up to D24. The average return, standard deviation, and variance are calculated as follows (Fig. 1.6):

Average yield:	=AVERAGE(D5:D24)
Standard deviation:	=STDEV.P(D5:D24)
Variance:	=VAR.P(D5:D24)

1.4 Assignment 4: Intrinsic Value Versus Time Value

Task

Show what the intrinsic value of a call option (put option) is if the Pharma Group share is at EUR 70 at maturity. Is the respective option in the money, at the money, or out of the money?

Content

In addition to intrinsic value, each option also has a time value. This is the amount that market participants are willing to pay in anticipation of certain market price changes. The time value can be calculated as the difference between the actual option price and the intrinsic value. The following applies: time value = option price – intrinsic value.

The time value of an option depends on various value drivers, in particular, volatility, the risk-free interest rate, expected future dividend yields, the remaining term of the option, and the current pricing. In the example of a call option, the time value reflects the chances for the holder of an option that the underlying asset will take a favorable course—i.e., will rise as much as possible. In the case of American options, the probability of a favorable trading day naturally increases as the term of the option increases. The longer the remaining term of the option, the greater the time value of American options.

The option price is calculated as follows:

Option price = intrinsic value + time value

The *intrinsic value* is the positive difference between the current price of the underlying asset and the strike price. It cannot be negative but can take the value zero.

The *fair value* is the amount that market participants are willing to pay in anticipation of market price changes. The time value decreases exponentially as the remaining term decreases. The shorter the period until the expiration date, the greater the risk of a worthless expiration.

A call or put option can be in the money (In The Money (ITM)), out of the money (OTM), or at the money (ATM). This is clearly illustrated in Fig. 1.7.

Important Formulas

1. Inner value of the call option: = max (S-X;0)

   IF(Pharma Group share price > exercise price; Pharma Group share price-exercise price; 0)

	In the money	At the money	Out of the money
Call	Underlying price > Strike price	Underlying price = Strike price	Underlying price < Strike price
Put	Underlying price < Strike price	Underlying price = Strike price	Underlying price > Strike price

Fig. 1.7 Possible pricing of call and put options

	A	B	C	D	E
1					
2		**Call or Put option on the share of Pharma Group**			
3					
4		**Closing price of the share on maturity date:**			70.00
5					
6		**Type**	**Strike Price**	**Intrinsic value**	**IN, AT or OUT of the money**
7		Call	68.00	2.00	**IN THE MONEY**
8		Put	72.00	2.00	**IN THE MONEY**
9					

Fig. 1.8 Calculation of the intrinsic value of a call or put option on the Pharma Group share

2. Inner value of the put option: = max (X - S;0)

IF(Pharma Group share price < exercise price; Pharma Group exercise price share price; 0)

3. Calculation "in the money, at the money, or out of the money":

=If($E$4>=C7;IF($E$4=C7; "AT THE MONEY"; "IN THE MONEY"); "OUT OF THE MONEY")

Display in Excel (Fig. 1.8)
Cell D7 is calculated as follows: =IF($E$4> C7;$E$4-C7;0)

Cell D8 is calculated as follows: =IF ($E$4<C8;C8-$E$4;0)

For example, cell E7 is calculated as follows: =IF($E$4>=C7;IF($E$4=C7; "AT THE MONEY"; "IN THE MONEY"); "OUT OF THE MONEY")

1.5 Assignment 5: The Binomial Model—The Single Period Case

Task
Calculate the call price according to the assumptions made in Assignment 2. These are essentially:

Price of the underlying:	Euro 70
Base price of the option:	Euro 68
Maturity of the option in years:	1
Implied volatility:	30%
Risk-free interest rate:	1%

It is based on a European call option.
These premises are assumed in the following unless otherwise mentioned.

Content
In the binomial model, the option price is valued when the underlying rises and when it falls. The financial model starts with a scenario t = 0 and ends with two scenarios t = 1. This is referred to as the single-period case. The binomial model is based in particular on the following premises:

1. No dividend payments during the term.
2. Perfect capital market (e.g., there are no transaction costs and taxes).
3. Any capital investment and borrowing at the risk-free interest r_f is possible.
4. Arbitrage profits are not possible.
5. Short sales are unrestricted.
6. Securities are divisible as desired.

Important Formulas
In the first step, the following are calculated:

- Gradient factor
- Reduction factor
- Pseudo probability
- The formula for the slope factor is: $u = e^{\sigma\sqrt{\Delta T}}$
- The formula for the sink factor is: $d = e^{-\sigma\sqrt{\Delta T}}$
- The formula for the pseudo-probability is: $q = \frac{(1+r_f)-d}{u-d}$
- The formula for the price of the call is: $C = \frac{q * C_0 + (1-q) * C_d}{1+r_f}$

The increase and decrease factors depend on the time step length and the amount of the standard deviation. The formulas shown are based on the so-called Wiener Process or Geometric Brownian Motion, which are used to model possible price changes of typical products traded in capital markets.

This is followed by the calculation of the value development tree. What value will the Pharma Group share price assume if the share price develops positively (current share price * increase factor) or negatively (current share price * decrease factor)? The third step is the calculation of the option price tree. The intrinsic value of the call option after 1 year is calculated using the formula from Assignment 2 for the positive case. Then, the formula is dragged down with the mouse. In the last step, to calculate the price of the call, the respective stock price is multiplied by the pseudo-probability or (1 – pseudo-probability) and discounted by the risk-free interest rate. The pseudo-probability is called this because it is not an estimated quantity but is calculated from the known quantities u, d, as well as r_f. It is obtained by solving the formula for the price of the call for q.

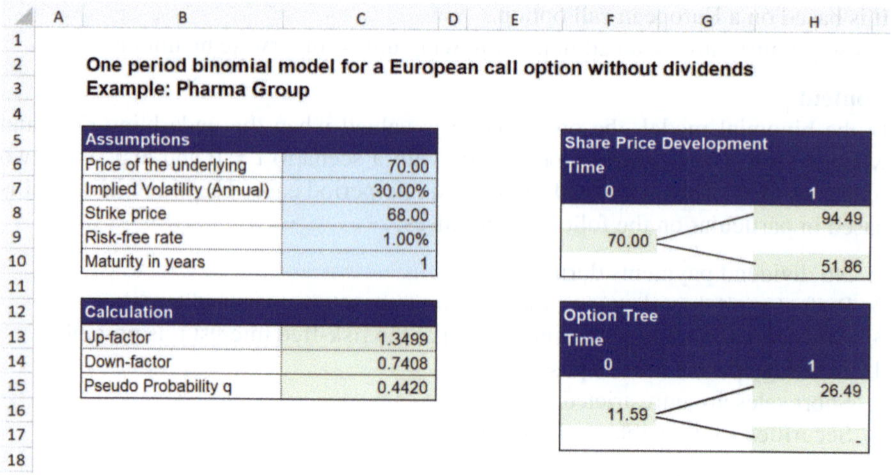

Fig. 1.9 Calculation of the call price in the context of the single-period case

Display in Excel

The assumptions are entered in cells C6–C10. The slope factor in C13 is calculated as follows: =EXP(C7*SQRT(C10)). Similarly, the slope factor in C14 is calculated as =EXP(-C7*SQRT(C10)). The pseudo-probability is calculated as =(1+C9-C14)/(C13-C14).

In the next step, the share price is calculated for the two cases: (1) increased share price (S_u) or (2) decreased share price (S_d). S_u in H8 is calculated from =F9*C13. S_d in H10 results from =F9*C14. Then the intrinsic value of the option is calculated for both cases. This leaves the value development tree and takes us to the option price tree. The intrinsic value results in H15 as =If(H8> C8;(H8-C8);0). H17 =IF(H10> C8;(H10-C8);0). Thereupon the expected value is formed as H15*C15+H17*(1-C15). This in turn is discounted by the risk-free interest rate r_f as shown above in the formula for the price of the call, and the price of the call in F16 is thus obtained as =H15*C15/(1+C9)+H17*(1-C15)/(1+C9) (Fig. 1.9).

Finally, premise 4 can be verified. If the following holds:

$d < (1+r_f) < u$, then no arbitrage is possible. This is given by $0.74 < 1.01 < 1.35$.

1.6 Assignment 6: The Multi-period Case—Six Steps (Bimonthly)

Task

Calculate the call price based on the assumptions of Assignment 4. However, the term is now divided into six 2-month periods, and a dividend yield of 4.85% is assumed. The dividend yield is assumed to accrue in the middle of the current year.

Content

The calculation for Assignment 6 is the same as for Assignment 5. However, the following three calculations change:

- Instead of the slope factor, the monthly slope factor is calculated.
- Instead of the reduction factor, the monthly reduction factor is calculated.
- Instead of the risk-free interest rate, the monthly interest rate is calculated.

Thus, a more elaborate value development tree must now be created. How can this be done efficiently?

Value development tree Link cell B28 with cell C6. Plot the stock price at rising prices (C27) and falling prices (C29). Now fix $C$14 and $C$15, respectively.

Next, take cell C29 and drag it to cell C28. Then, take cell C28 and drag it to cell H28. Then, take cell C27 and drag it to H27. Again, drag cell H27 to H22. Continue in the same way with the lower part: C29 to H29 and H29 to H34.

Subsequently, you build in the dividend yield. We start from the premise that the dividend yield accrues in the middle of the year. Accordingly, click on E25 and multiply this term by (1-dividend yield), fixing the dividend yield. You now drag cell E25 to cell E29, after which you click cell E31 and multiply it by the term (1-dividend yield). The value development tree is finished. You can now format it correctly by selecting it and formatting it with the right mouse button. Enter the following into the field: _-* #,##0.00_-;[Red](#,##0.00);_-* "-"_-;_-@_-

Option price tree To create the option price tree, you only need to redefine two cells with formulas.

1. *Cell:* Cell H42 is used to store the formula from Assignment 4, with the Strike Price fixed in dollars. Then, take cell H42 and drag it to cell H54.
2. *Cell:* In cell G43, store the formula for the price of the call from Assignment 5: =H42*$C$17/(1+$C$16)+H44*(1-$C$17)/(1+$C$16), fixing the risk-free rate and the pseudo-probability in dollars. Then, take cell G43 and drag it to cell G53. Now, drag the cells thus marked in column G to column B altogether. The option price tree is finished. You can now format it in the same way as the value development tree. The result is €7.50 for the call option.

Important Formulas

The following three formulas change:

fx

$$u_M = e^{\sigma\sqrt{\frac{1}{12}}}$$

u_M = up = monthly slope factor

$$d_M = e^{-\sigma\sqrt{\frac{1}{12}}}$$

d_M = down = monthly decrease factor

$$r_{fM} = \sqrt[12]{\left(1+r_f\right)}-1$$

r_{fM} = monthly interest

Multi-period binomial model (6 steps) for a European call option with dividends
Example: Pharma Group

Assumptions	
Price of the underlying	70.00
Implied Volatility (Annual)	30.00%
Strike price	68.00
Risk-free rate	1.00%
Dividend	4.85%
Maturity in months	12

Calculation	
Up-factor (per step)	1.1303
Down-factor (per step)	0.8847
Risk-free rate (per step)	0.0017
Pseudo Probability q	0.4762

Share Price Development

Time 0	1	2	3	4	5	6
						138.88
					122.87	
				108.71		108.71
			96.18		96.18	
		89.43		85.09		85.09
	79.12		75.28		75.28	
70.00		70.00		66.61		66.61
	61.93		58.93		58.93	
		54.79		52.13		52.13
			46.13		46.13	
				40.81		40.81
					36.10	
						31.94

Option Tree

Time 0	1	2	3	4	5	6
						70.88
					54.99	
				40.93		40.71
			28.72		28.29	
		19.11		17.70		17.09
	12.18		10.43		8.13	
7.50		5.92		3.86		-
	3.27		1.84		-	
		0.87		-		-
			-		-	
				-		-
					-	
						-

Fig. 1.10 Calculation of the call price in the multi-period case (6 steps)

Display in Excel

In Excel, this results in the following formulas (Fig. 1.10):

- u_M => C14 =EXP(C7*SQRT(2/C11))
- d_M => C15 =EXP(-C7*SQRT(2/C11))
- r_{fM} => C16 =((1+C9)^(2/C11)-1)

1.7 Assignment 7: The Multi-period Case—12 Steps (1 Month) European Call

Task
Calculate the call price based on the assumptions of Assignment 4. However, the term is now divided into 12 monthly phases. A dividend yield of 4.85% is still assumed to accrue at mid-year.

Content
The running time of the following three formulas changes in the sense that the fraction is now not 2/12 but 1/12:

- u_M = up= monthly slope factor
- d_M = down= monthly decrease factor
- r_{fM} = monthly interest

Important Formulas
In Excel, this results in the following formulas:

- u_M => C14 =EXP(C7*SQRT(1/C11))
- d_M => C15 =EXP(-C7*SQRT(1/C11))
- r_{fM} => C16 =((1+C9)^(1/C11)-1)

Display in Excel
The calculation is analogous to the calculation of Assignment 5. As a result, you get €7.60 for the call option (Fig. 1.11).

1.8 Assignment 8: The Multi-period Case—12 Steps (1 Month), American Call

Task
Calculate the call price based on the assumptions of Assignment 6. However, this assignment assumes an American call option rather than a European call option.

Content
In contrast to Assignment 6, an *American* call option with 12 steps and a dividend yield of 4.85% is now assumed. The value development tree remains unchanged. When creating the option price tree, note the significant change in cell M55.

In contrast to European options, which permit exercise only once at the end of the term, American options can be exercised during the entire term. This may result in a different option price than for European options, depending on the situation of the various price factors. Accordingly, an adjustment of the price calculation model is also necessary to take into account the possibility of early exercise at each step of the model.

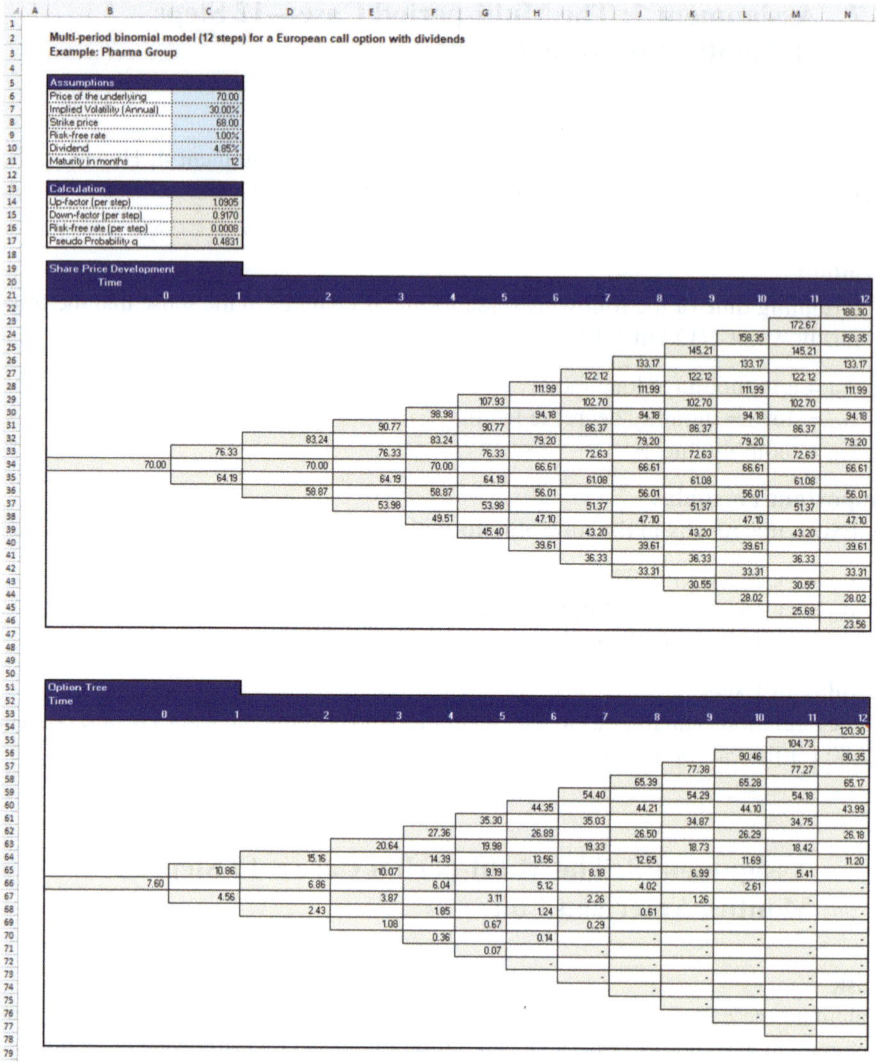

Fig. 1.11 Calculation of the call price (12 steps) European call

Important Formulas

When calculating the option price tree at time 12, there is still no difference, but the formula for times 11 to 0 is different. In cell M55, the Financial Engineer stores a What-if function, which now additionally checks at each step whether an earlier exercise of the option would be profitable. This is the case if the calculated option value from the binomial model is lower than the intrinsic value in the period. Cell M55 is now drawn to cell M77 and then to column B, as discussed above. The result for the American call with a dividend yield and 12 steps is €8.11.

The formula for cell M55 is as follows:

=IF((N54*$C$17/(1+$C$16)+N56*(1-$C$17)/(1+$C$16))<(M23-$C$8);M23-$C$8;N54*$C$17/(1+$C$16)+N56*(1-$C$17)/(1+$C$16))

Display in Excel (Fig. 1.12)

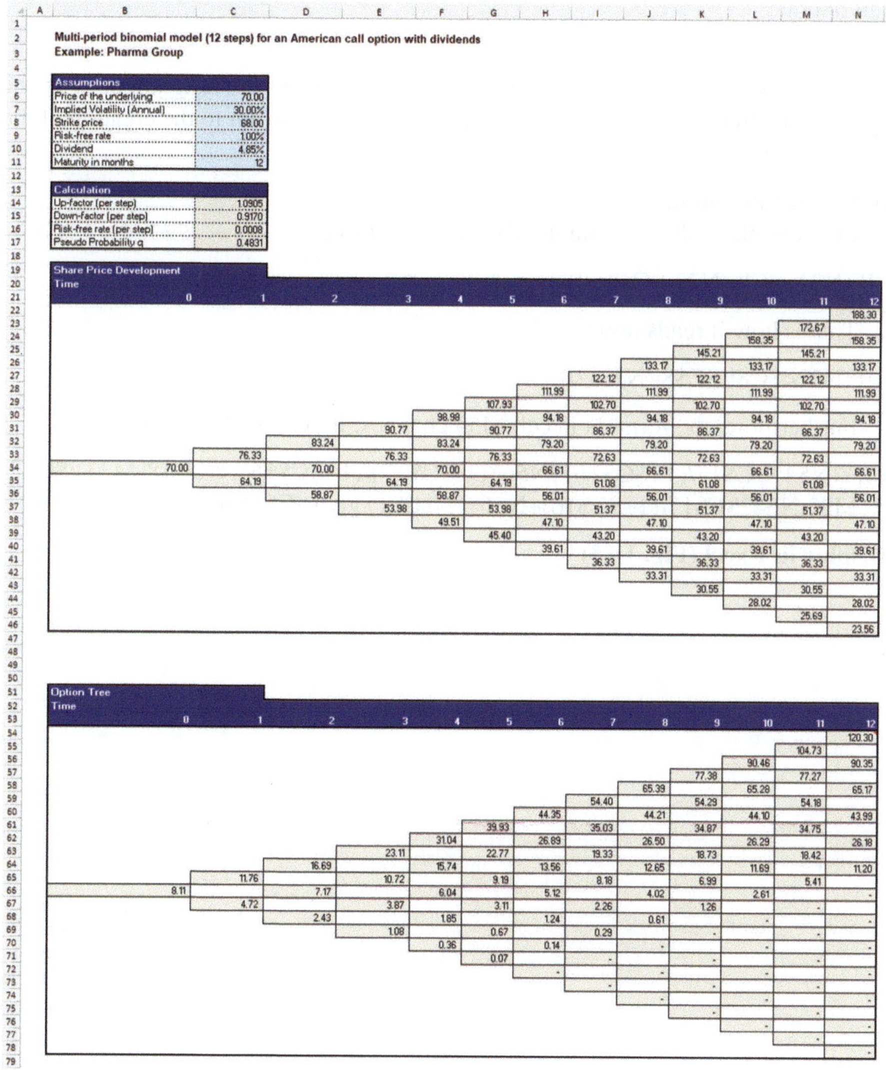

Fig. 1.12 Calculation of the call price (12 steps) American call

1.9 Assignment 9: The Multi-period Case—12 Steps (1 Month), American Put

Task
Assignment 8: Calculate the put price based on the assumptions of assignment 18. However, this assignment assumes an American put option rather than an American call option.

Content
The value development tree remains the same compared to Assignment 7. In the option price tree, there is a difference in the two cells relevant to the creation of the option price tree.

Important Formulas
While exemplary, the formula in cell N54 was before

=IF(N22>$C8;(N22-$C$8);0)

That is how it reads now:

=IF(N22<$C8;($C$8- N22);0).

Furthermore, the formula in cell M55 changes. It now reads as follows:

=IF((N54*$C$17/(1+$C$16)+N56*(1-$C$17)/(1+$C$16))<($C$8-M23);$C$8-M23;N54*$C$17/(1+$C$16)+N56*(1-$C$17)/(1+$C$16))

Display in Excel (Fig. 1.13)

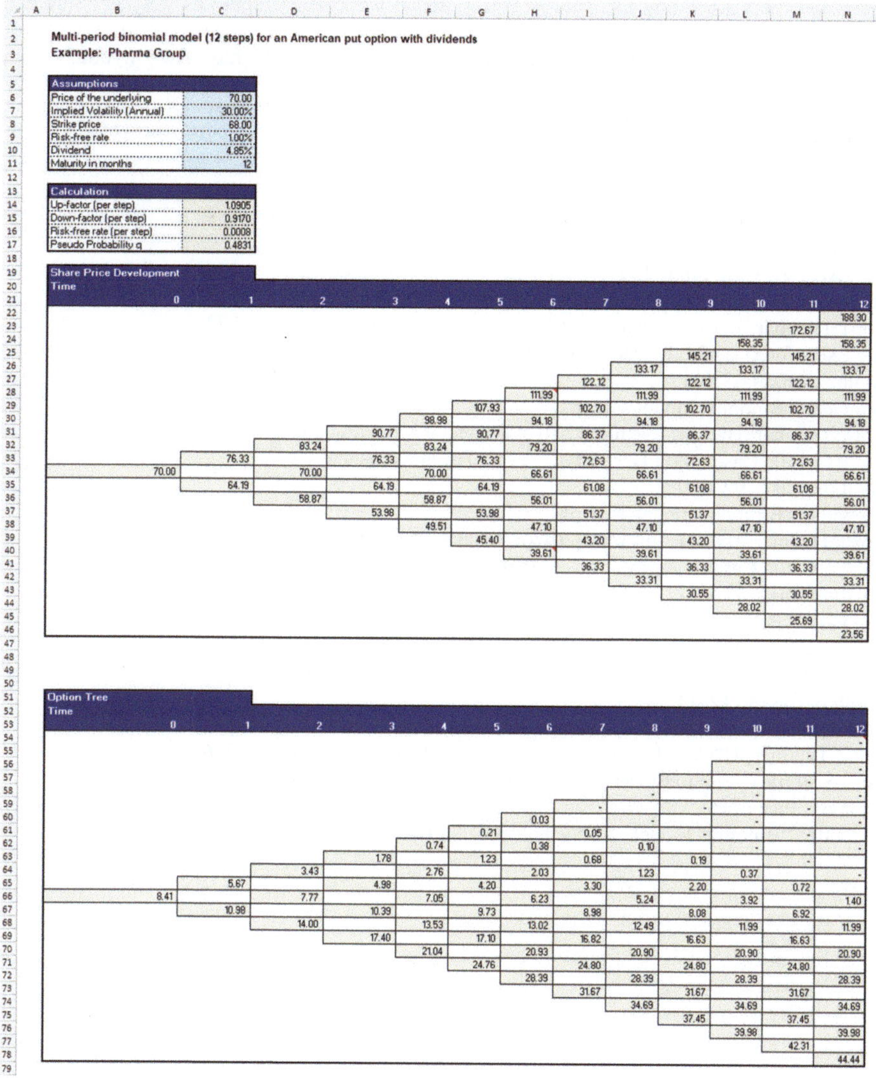

Fig. 1.13 Calculation of the put price (12 steps) American put

Further Reading

Ernst, Dietmar; Häcker, Joachim (2016): Financial Modeling, pp. 751–756.

Bloss, M.; Ernst, D.; Häcker, J.; Sörensen, D. (2010): Financial Engineering, pp. 89–93.

FAZ of 25.3.2008: Historical financial crises: Netherlands 1637: A bulb for 87,000 euros.

Toolbox, tasks 1–15.

See video "An Introduction to Options"

Ernst, Dietmar; Häcker, Joachim (2016): Financial Modeling, pp. 17–45.

See Excel file Case Study Derivatives Part 1, Excel worksheet "Input" as well as "Summary Option."
See video "How to avoid mistakes in Excel models" https://www.youtube.com/watch?v=mtpAV3YhHgA
Ernst, Dietmar; Häcker, Joachim (2016): Financial Modeling, pp. 756–765 and pp. 608–622.
See Excel file Case Study Derivatives Part 1, Excel worksheet 6.
Toolbox, Tasks 16–35.
See video "Value Drivers and Basics of Option Pricing."
Ernst, Dietmar; Häcker, Joachim (2016): Financial Modeling, pp. 765–768.
Toolbox, Tasks 36–63.
See Excel file Case Study Derivatives Part 1, Excel worksheet 8.
See video "Value Drivers and Basics of Option Pricing."
Ernst, Dietmar; Häcker, Joachim (2016): Financial Modeling, pp. 773–777.
See Excel file Case Study Derivatives Part 1, Excel worksheet 9.
Toolbox, Tasks 67–91.
See video "Binomial model–1 step."
Ernst, Dietmar; Häcker, Joachim (2016): Financial Modeling, pp. 778–780.
See Excel file Case Study Derivatives Part 1, Excel worksheet 10.
Toolbox, Tasks 92–109.
See video "Binomial model–6 steps."
Ernst, Dietmar; Häcker, Joachim (2016): Financial Modeling, pp. 780–783.
See Excel file Case Study Derivatives Part 1, Excel worksheet 11.
Toolbox, Tasks 110–111.
See video "Binomial model - 12 steps."
Ernst, Dietmar; Häcker, Joachim (2016): Financial Modeling, pp. 783–784.
See Excel file Case Study Derivatives Part 1, Excel worksheet 12.
Toolbox, Tasks 112–115.
See video "Binomial model - 12 Steps American Call."
Ernst, Dietmar; Häcker, Joachim (2016): Financial Modeling, pp. 785–786.
See Excel file Case Study Derivatives Part 1, Excel worksheet 13.
Toolbox, Tasks 116.
See video "Binomial model - 12 Steps American Put."

Chapter 2
The Black-Scholes Model, Including the Greeks

2.1 Assignment 10: Pricing Options with the Black-Scholes Model

Task
Calculate the price of the call and the put without dividend yield according to the Black-Scholes model based on the premises of Chap. 1. Explain how the two results are related and how the implied volatility can be calculated in the case of Pharma Group.

Content
The Black-Scholes model is the dominant model for valuing options. In principle, the Black-Scholes model is a special case of the binomial model. When the binomial model is made constant (the number of steps is always increased), the results of the binomial model converge to the results of the Black-Scholes model.

The Black-Scholes model is essentially based on the following 12 premises:

1. *Rational investors:* Investors behave rationally.
2. *Perfect capital market:* The capital market can be regarded as perfect: Information is equally accessible to all investors. There are no barriers to entry that would prevent market participants from gaining free access to all markets.
3. *No taxes:* No taxes exist.
4. *Log-normal distribution function:* The probability distribution of potential stock prices is a log-normal distribution.
5. *No subscription rights, dividends, or share splits:* No dividends will be paid on the underlying share during the term of the option. Furthermore, there is no stock split, and no subscription rights are granted.

Supplementary Information The online version contains supplementary material available at https://doi.org/10.1007/978-3-031-85822-2_2.

D. Ernst, J. Häcker, *Derivatives*, Springer Texts in Business and Economics,
https://doi.org/10.1007/978-3-031-85822-2_2

6. *European options:* All options treated can only be exercised at maturity and are therefore European options.
7. *No arbitrage profits:* The premise is that if there is an immediate risk-free profit due to mispricing, it will disappear immediately due to an adjustment in supply and demand.
8. *No transaction and information costs:* There are no transaction and information costs.
9. *Short sales*: Short sales are possible without limit.
10. *Known and constant volatility:* The volatility of the underlying price remains constant during the term and is known in advance.
11. *Arbitrary divisibility:* All assets and derivatives are arbitrarily divisible and do not react in price to changing liquidity.
12. *Risk-free interest:* Investments and loans are both possible at risk-free interest for an unlimited amount. The risk-free interest rate remains constant and known during the term.

There is a mathematical relationship between the prices of a call and a put—the so-called put-call parity. In essence, this means that a portfolio can be hedged in two different ways, with both types leading to the same result.

1. *Synthetic hedge:* The financial engineer forms a portfolio of a risk-free investment and calls on an underlying asset—in this case, the Pharma Group share. The former represents the price floor. The latter allows the financial engineer to participate in a possible positive price trend of the underlying.
2. *Protective Put:* The Financial Engineer hedges his portfolio consisting of the Pharma Group share (S_0) with put options (p). The maturity and the strike price of the call and the put are the same. The payoff profile is the same as for the synthetic hedge.

In other words, the price of the call plus the cash (including the steady interest) needed to buy the Pharma Group share is equal to the price of the put plus the Pharma Group share price.

Important Formulas
The value of a call option is calculated according to the Black-Scholes model as follows:

fx $$c = s_0 N(d_1) - Xe^{-r_f T} N(d_2)$$

with

$$d_1 = \frac{\ln\left(\frac{s_0}{X}\right) + \left(r_f + \frac{\sigma^2}{2}\right)T}{\sigma\sqrt{T}}$$

$$d_2 = d_1 - \sigma\sqrt{T}$$

The value of a put option is calculated as follows:

$$p = Xe^{-r_f T} N(-d_2) - S_0 N(-d_1)$$

The value of the protective put (left side of the following equation) is equal to the value of the synthetic hedge (right side of the following equation). The formula for the put-call parity is:

$$p + S_0 = c + X \times e^{-r_f \times T}$$

The left-hand side consists of the price of the put and the share price at time t_0. The right-hand side of the equation corresponds to the price of the call and the value of the fixed-interest investment in the amount of the strike price (X) of the call, which is now discounted by the remaining term (T).

Resolved for p, the price of a put results according to the put-call parity.

$$p = c + X \times e^{-r_f \times T} - S_0$$

The value of the fixed-interest investment in the amount of the strike price (X) of the call option, which is now discounted by the remaining term (t). The risk-free interest rate is denoted by "r," the Eulerian number by "e." Thus:

Display in Excel

d_1 => cell C13	=(LN(C6/C7)+(C9+0,5*C10^2)*C8)/(C10*SQRT(C8))
d_2 => cell C14	=C13-ROOT(C8)*C10
$N(d_1)$ => cell C15	=NORM.S.DIST(C13; TRUE)
$N(d_2)$ => cell C16	=NORM.S.DIST(C14; TRUE)
Call price => cell C19	=C6*C15-C7*EXP(-C9*C8)*C16
Put price => cell C20	=C7*EXP(-C9*C8)*(1-C16)-C6*(1-C15)

The call price of EUR 9.59 and the put price of EUR 6.92 refer to European call options without dividend yield.

With the put-call parity, the financial engineer can constantly check that his calculation does not contain an error. For this purpose, the control calculation cell in C25 must constantly show 0. The value of the protective put (C24) must always be equal to the value of the synthetic hedge (C23) (Fig. 2.1).

Black Scholes Model (without dividends)

Example: Pharma Group

Assumptions	
Price of the underlying	70.00
Strike price	68.00
Maturity (in years)	1
Risk-free rate	1.00%
Implied Volatility	30.00%

Calculations	
d_1	0.2800
d_2	(0.0200)
$N(d_1)$	0.6102
$N(d_2)$	0.4920

Result	
Call Price	9.59
Put Price	6.92

Control Calculation	
c+X * exp(-rf * T)	76.92
p+S	76.92
Difference	-

Fig. 2.1 Calculation of the call price without dividend yield according to the Black-Scholes model

2.2 Assignment 11: Pricing Options with the Black-Scholes-Merton Model

Task

Calculate the call and put price with dividend yield according to the Black-Scholes-Merton model based on the premises of Chap. 1 (plus the dividend yield). Compare their result with the results obtained based on the application of the binomial model. Furthermore, comment on the effect of the dividend yield on the option price.

Content

In the Black-Scholes-Merton model, the Black-Scholes model is extended by including the dividend yield. In the present case, in contrast to the Black-Scholes model, a dividend yield of 4.85% is now included. The results of this model come closest to the actual market prices.

The implied volatility was shown in assignment 2. The implied volatility can be derived from the currently traded option on Pharma Group using the Black-Scholes-Merton formula and the iteration method.

In the present case, it becomes clear that, using the European call option including dividend yield as an example, the binomial model converges more and more in the direction of the Black-Scholes-Merton model with an increasing number of steps or smaller step size (see Fig. 1.5). With the multi-period binomial model with six steps (Assignment 6), a price of 7.50 is obtained. Doubling the steps to 12 steps (Assignment 7) already leads to a price of 7.60 euros. The continuous Black-Scholes-Merton model yields a price of 7.67 euros. Thus, the results of the binomial model, with an increasing number of steps, approach the results of the Black-Scholes-Merton model. The discrete evolution thus approximates the continuous evolution. This approximation occurs due to the up- and down-development in the binomial model from both sides (over- and underestimation). The background is the deviation between the binomial model and the Black-Scholes-Merton model in the expected value of close to zero, which, however, is distributed in both directions. The variance (distribution of the price development) is somewhat coarser (in both directions) in the binomial model than the premise of normal distribution in the Black-Scholes-Merton model.

It is also clear that dividend payments have a direct impact on the price of the underlying. Dividend payments cause the price of the underlying to fall. Due to the direct influence, dividend payments lead to falling prices in the case of call options and rising prices in the case of put options. Based on the Black-Scholes model, the price of a call option decreases from 9.59 euros (without dividend yield) (Assignment 10) to 7.67 euros (with a dividend yield of 4.85%) (Assignment 11). While for the European call with dividend yield, the binomial model approaches the Black-Scholes-Merton model "from below" with an increasing number of steps, this happens "from above" for the American call.

Important Formulas

This changes the first part of the formula for calculating a call option by adding the dividend yield (e^{-DT}) to this term.

The value of a call option is now calculated as follows:

$$fx \quad c = S_0 \times e^{-D\times T} \times N(d_1) - X \times e^{-r_f\times T} \times N(d_2)$$

with

$$d_1 = \frac{\ln\left(\frac{S_0}{X}\right) + \left(r_f - D + \frac{\sigma^2}{2}\right)\times T}{\sigma \times \sqrt{T}}$$

$$d_2 = d_1 - \sigma \times \sqrt{T}$$

The value of a put option is calculated as follows:

$$p = -S_0 \times e^{-D\times T} \times N(-d_1) - X \times e^{-r_f\times T} \times N(-d_2)$$

In the Black-Scholes-Merton model, the put-call parity is extended by the dividend yield and is calculated as follows:

$$p + S_0 = c + X \times e^{-r_f\times T} + D$$

Display in Excel

The formulas listed below are relevant to the calculation of the Greeks presented in the following assignments, and therefore, the call and put formulas are broken down into smaller units.

Exp(-rf *T) => cell C14	=EXP(-C11*C9)
Exp(-DT) => cell C15	=EXP(-C8*C9)
(rf-D+0.5*V2) => cell C16	=(C11-C8+0.5*C10^2)
d_1 => cell C17	=(LN(C6/C7)+C16*C9)/(C10*SQRT(C9))
$N(d_1)$ => cell C18	=NORM.S.DIST(C17; TRUE)
$N(-d_1)$ => Cell C19	=1-C18
d_2 => cell C20	=C17-(C10*SQRT(C9))
$N(d_2)$ => cell C21	=NORM.S.DIST(C20; TRUE)
$N(-d_2)$ => cell C22	=1-C21

Price of the option (long call) => cell C25 = C6*C15*C18-C7*C14*C21

Price of the option (long put) => cell C26 = -C6*C15*C19+C7*C14*C22

Black Scholes Merton Model (including dividends)
Example: Pharma Group

Assumptions	
Price of the underlying	70.00
Strike price	68.00
Dividend	4.85%
Maturity in years	1.00
Implied Volatility (Annual)	30.00%
Risk-free rate	1.00%

Calculation	
Exp(-rf *T)	0.9900
Exp(-DT)	0.9527
(rf-D+0.5*V^2)	0.0065
d1	0.1183
N (d1)	0.5471
N (-d1)	0.4529
d2	-0.1817
N(d2)	0.4279
N(-d2)	0.5721

Result	
Call price	7.67
Put price	8.31

Fig. 2.2 Calculation of the call price with dividend yield according to the Black-Scholes-Merton model

The implied volatility in cell C10 can be determined for a given option price (here, for example, the call of EUR 7.67) by the Financial Engineer clicking on "Edit/Target Search" and then entering the following (Fig. 2.2):

Target cell => C25
Target value => 7.67 (with all decimal places)
Changeable cell => C10

After pressing "OK," Excel iterates 100 times according to the default setting, and you get the result: 30.00%.

2.3 Assignment 12: Greeks—Delta

Task
Calculate the delta for a call option and a put option on Pharma Group stock based on the premises of Assignment 5.

Content
In mathematical terms, the delta is the first partial derivative of the option price according to the price of the underlying. The delta indicates, as sensitivity, how the option price changes when the underlying changes.

The following applies to the values of Delta (Fig. 2.3):

Important Formulas
The formulas for calculating the delta of a call option and a put option are:

$$Delta\ Call = e^{-D*T} * N\left(d_1\right)$$

$$Delta\ Put = e^{-D*T} * \left(N\left(d_1\right) - 1\right)$$

Display in Excel

Delta (long call) => cell C27 =C15*C18
Delta (long put) => cell D27 =C15*(C18-1)

The delta (long call) is 0.52. This means that the call rises (falls) by 0.52 euros if the Pharma Group share rises (falls) by 1 euro (Fig. 2.4).

Delta of	Out of the money	At the money	In the money
Long call / short put	Approx. 0 to 0.5	Approx. 0.5	Approx. 0.5 to 1
Long put / short call	Approx. 0 to -0.5	Approx. -0.5	Approx. -0.5 to -1

Fig. 2.3 Values of the delta

Black Scholes Merton Model (including the derivation of the Greeks)
Example: Pharma Group

Assumptions	
Price of the underlying	70.00
Strike price	68.00
Dividend	4.85%
Maturity in years	1.00
Implied Volatility (Annual)	30.00%
Risk-free rate	1.00%

Calculations	
Exp(-rf *T)	0.9900
Exp(-DT)	0.9527
(rf-D+0.5*V²)	0.0065
d1	0.1183
N(d1)	0.5471
N(-d1)	0.4529
d2	-0.1817
N(d2)	0.4279
N(-d2)	0.5721
Probability density function of the Standard Normal Distribution	0.3962

Result	Long Call	Long Put
Option Price	7.67	8.31
Delta	0.5212	-0.4315

Fig. 2.4 Calculation of the delta

2.4 Assignment 13: Greeks—Gamma

Task

Calculate the delta for a call option and a put option on Pharma Group stock based on the premises of Assignment 5.

Content

The change in delta is described by gamma. Gamma indicates how delta changes when the price of the underlying changes.

Important Formulas

The formula for calculating the gamma of a call option and a put option is the same and is as follows:

Black Scholes Merton Model (including the derivation of the Greeks)
Example: Pharma Group

Assumptions	
Price of the underlying	70.00
Strike price	68.00
Dividend	4.85%
Maturity in years	1.00
Implied Volatility (Annual)	30.00%
Risk-free rate	1.00%

Calculations	
Exp(-rf *T)	0.9900
Exp(-DT)	0.9527
(rf-D+0.5*V²)	0.0065
d1	0.1183
N(d1)	0.5471
N(-d1)	0.4529
d2	-0.1817
N(d2)	0.4279
N(-d2)	0.5721
Probability density function of the Standard Normal Distribution	0.3962

Result	Long Call	Long Put
Option Price	7.67	8.31
Gamma	0.0180	0.0180

Fig. 2.5 Calculation of the gamma

$$Gamma\ Call\ and\ Put = \frac{\left(\frac{1}{\sqrt{2\pi}}\right) * e^{-\frac{d_1^2}{2}} * e^{-D*T}}{S_0 * \sigma * \sqrt{T}}$$

Display in Excel

The probability density function of the standard normal distribution => cell C23 =EXP(-(C17^2)/2)/ROOT(2*PI())

Gamma (long call) => cell C28 =C23*C15/(C6*C10*SQRT(C9))
Gamma (long put) => cell D28 =C23*C15/(C6*C10*SQRT(C9))

The gamma (long call and long put) is 0.018. This means that for the call and put options, the delta of the option increases (or decreases) by 0.02 euros if the Pharma Group share increases (or decreases) by 1 euro (Fig. 2.5).

2.5 Assignment 14: Greeks—Theta

Task
Calculate the theta for a call option and a put option on Pharma Group stock based on the premises of Assignment 5.

Content
The theta indicates how the option price changes when a day's worth of time is lost. Since the option price formula refers to the term of the option in years, the result must be converted to 1 day. Here, 250 trading days are assumed.

Important Formulas
The formula for calculating the theta of a call option is:

$$\textit{Theta Call} = \left\{ -\frac{S_0\left(\frac{1}{\sqrt{2\pi}}\right)e^{-\frac{d_1^2}{2}}\sigma e^{-DT}}{2\sqrt{T}} + \mathrm{D}S_0N(d_1)e^{-DT} - r_f\,Xe^{-r_fT}N(d_2) \right\} / 250$$

The formula for calculating the theta of a put option is:

$$\textit{Theta Put} = \left\{ -\frac{S_0\left(\frac{1}{\sqrt{2\pi}}\right)e^{-\frac{d_1^2}{2}}\sigma e^{-DT}}{2\sqrt{T}} - \mathrm{D}S_0N(-d_1)e^{-DT} + r_f\,Xe^{-r_fT}N(-d_2) \right\} / 250$$

Display in Excel
Theta (long call) => cell C29
=(-((C6*C23*C10*C15)/(2*SQRT(C9)))+(C8*C6*C18*C15)-(C11*C7*C14*C21))/250

Theta (long put) => cell D29
=(-((C6*C23*C10*C15)/(2*SQRT(C9)))-(C8*C6*C19*C15)+(C11*C7*C14*C22))/250

The theta (long call) is –0.0099. For the call option, this means that the call falls by just under 1 cent overnight (Fig. 2.6).

	A	B	C	D	E
4					
5		**Assumptions**			
6		Price of the underlying	70.00		
7		Strike price	68.00		
8		Dividend	4.85%		
9		Maturity in years	1.00		
10		Implied Volatility (Annual)	30.00%		
11		Risk-free rate	1.00%		
12					
13		**Calculations**			
14		Exp(-rf *T)	0.9900		
15		Exp(-DT)	0.9527		
16		(rf-D+0.5*V²)	0.0065		
17		d1	0.1183		
18		N(d1)	0.5471		
19		N(-d1)	0.4529		
20		d2	-0.1817		
21		N(d2)	0.4279		
22		N(-d2)	0.5721		
23		Probability density function of the Standard Normal Distribution	0.3962		
24					
25		**Result**	**Long Call**	**Long Put**	
26		Option Price	7.67	8.31	
29		Theta	-0.0099	-0.0202	

Fig. 2.6 Calculation of the theta

2.6 Assignment 15: Greeks—Rho

Task

Calculate the Rho for a call option and a put option on Pharma Group stock based on the premises of Assignment 5.

Content

The Rho indicates how the option price changes when the interest rate changes.

Important Formulas

The Rho of a call option is calculated as follows:

$$Rho\,Call = X * T * e^{-r_f * T} * N(d_2)$$

The Rho of a put option is calculated as:

$$Rho\,Put = -X * T * e^{-r_f * T} * N(-d_2)$$

	A	B	C	D	E
1					
2		**Black Scholes Merton Model (including the derivation of the Greeks)**			
3		**Example: Pharma Group**			
4					
5		**Assumptions**			
6		Price of the underlying	70.00		
7		Strike price	68.00		
8		Dividend	4.85%		
9		Maturity in years	1.00		
10		Implied Volatility (Annual)	30.00%		
11		Risk-free rate	1.00%		
12					
13		**Calculations**			
14		Exp(-rf *T)	0.9900		
15		Exp(-DT)	0.9527		
16		(rf-D+0.5*V²)	0.0065		
17		d1	0.1183		
18		N(d1)	0.5471		
19		N(-d1)	0.4529		
20		d2	-0.1817		
21		N(d2)	0.4279		
22		N(-d2)	0.5721		
23		Probability density function of the Standard Normal Distribution	0.3962		
24					
25		**Result**	**Long Call**	**Long Put**	
26		Option Price	7.67	8.31	
30		Rho	28.8081	-38.5153	

Fig. 2.7 Calculation of the Rho

Display in Excel

Rho (long call) => cell C30 =C7*C9*C14*C21
Rho (long put) => cell D30 =-C7*C9*C14*C22

The Rho (long call) is 28.8081. This means that the call will increase by 29 cents if the risk-free interest rate increases by 1% in absolute terms (it increases from 1% to 2%) (Fig. 2.7).

2.7 Assignment 16: Greeks—Vega

Task

Calculate the Vega for a call option and a put option on Pharma Group stock based on the premises of Assignment 5.

Content
The Vega indicates how the option price changes when the volatility changes. This is the implied volatility and not the historical volatility.

Important Formulas
The Vega of a call and an option is calculated as follows:

$$Vega\ Call\,und\,Put = S_0 * \sqrt{T} * \left(\frac{1}{\sqrt{2\pi}}\right) * e^{-\frac{d_1^2}{2}} * e^{-D*T}$$

Display in Excel

Vega (long call) => cell C31 =C6*SQRT(C9)*C23*C15
Vega (long put) => cell D31 =C6*SQRT(C9)*C23*C15

The Vega (long call and long put) is 26.4184. This means, for example, that the call increases by 26 cents if the implied volatility increases by 1% in absolute terms, i.e., from 30% to 31% (Fig. 2.8).

	A	B	C	D	E
1					
2		**Black Scholes Merton Model (including the derivation of the Greeks)**			
3		**Example: Pharma Group**			
4					
5		**Assumptions**			
6		Price of the underlying	70.00		
7		Strike price	68.00		
8		Dividend	4.85%		
9		Maturity in years	1.00		
10		Implied Volatility (Annual)	30.00%		
11		Risk-free rate	1.00%		
12					
13		**Calculations**			
14		Exp(-rf *T)	0.9900		
15		Exp(-DT)	0.9527		
16		(rf-D+0.5*V^2)	0.0065		
17		d1	0.1183		
18		N(d1)	0.5471		
19		N(-d1)	0.4529		
20		d2	-0.1817		
21		N(d2)	0.4279		
22		N(-d2)	0.5721		
23		Probability density function of the Standard Normal Distribution	0.3962		
24					
25		**Result**	**Long Call**	**Long Put**	
26		Option Price	7.67	8.31	
31		Vega	26.4184	26.4184	

Fig. 2.8 Calculation of the Vega

2.8 Assignment 17: Leverage

Task
Calculate the leverage for a call option and a put option on Pharma Group stock based on the premises of Assignment 5.

Content
Although, strictly speaking, simple leverage is not a Greek at all, it is accorded a high status, especially among private investors. Put simply, it expresses how many warrants can be obtained for the price of one unit of the underlying with the same capital investment. However, this information is only relevant to a limited extent, as generally less capital is invested in warrants compared to shares. The statement of the simple leverage is therefore very limited.

The formula must take into account any subscription ratio (ratio) of the warrant that differs from 1:1. For the following calculations, we assume a standard warrant with a subscription ratio of 1:1.

Important Formulas
The leverage of a call option is calculated as follows:

$$Hebel\ Call = \left(\frac{S_0 * Ratio}{c} \right)$$

The leverage of a put option is calculated as:

$$Hebel\ Put = \left(\frac{S_0 * Ratio}{p} \right)$$

Display in Excel

Leverage (long call) => Cell C32 =C6/C26
Leverage (long put) => Cell D32 =C6/D26

The leverage (long call) is 9.1210. This means, for example, that for a call warrant, the investor only has to invest about 11% of what a direct investor has to spend (Fig. 2.9).

2.9 Assignment 18: Greeks—Omega

Task
Calculate the omega for a call option and a put option on the Pharma Group stock based on the premises of Assignment 5. Furthermore, show the signs of the Greeks and the other ratios, leverage and omega, for the four basic strategies (long call, short call, long put, short put) of options (see Chap. 4) and explain.

Black Scholes Merton Model (including the derivation of the Greeks)
Example: Pharma Group

Assumptions	
Price of the underlying	70.00
Strike price	68.00
Dividend	4.85%
Maturity in years	1.00
Implied Volatility (Annual)	30.00%
Risk-free rate	1.00%

Calculations	
Exp(-rf *T)	0.9900
Exp(-DT)	0.9527
(rf-D+0.5*V²)	0.0065
d1	0.1183
N(d1)	0.5471
N(-d1)	0.4529
d2	-0.1817
N(d2)	0.4279
N(-d2)	0.5721
Probability density function of the Standard Normal Distribution	0.3962

Result	Long Call	Long Put
Option Price	7.67	8.31
Leverage Factor	9.1210	8.4216

Fig. 2.9 Calculation of leverage

Content

The simple leverage alone, as just shown, has only low informative value about the real leverage of a warrant in most maturity stages. Therefore, an additional parameter, the delta, is necessary to show how the call or put price changes when the price of the underlying changes.

The Omega—also called Theoretical Leverage—now measures the percentage change in the price of the warrant depending on the percentage change in the price of the underlying asset. The Omega allows for a more realistic assessment of the leverage than the Simple Leverage, especially for warrants that are not heavily in-the-money.

Important Formulas

The formula for calculating the omega of a call option is:

$$Omega\ Call = e^{-D*T} * N\left(d_1\right) * \left(\frac{S_0 * Ratio}{c}\right)$$

The omega of a put option is calculated as:

$$Omega\,Put = e^{-D*T} * \left(N\left(d_1\right)-1\right) * \left(\frac{S_0 * Ratio}{p}\right)$$

This shows that the omega is obtained by multiplying the delta by the leverage.

Display in Excel

Omega (long call) => cell C33 =C15*C18*(C6/C26)
Omega (long put) => cell D33 =(C15*(C18-1)*(C6/D26))

The leverage (long call) is 4.7537. This means, for example, that for a call warrant, the call price increases by 4.75% if the Pharma Group share price increases by 1% (Fig. 2.10).

	A	B	C	D	E
1					
2		**Black Scholes Merton Model (including the derivation of the Greeks)**			
3		**Example: Pharma Group**			
4					
5		Assumptions			
6		Price of the underlying	70.00		
7		Strike price	68.00		
8		Dividend	4.85%		
9		Maturity in years	1.00		
10		Implied Volatility (Annual)	30.00%		
11		Risk-free rate	1.00%		
12					
13		Calculations			
14		Exp(-rf *T)	0.9900		
15		Exp(-DT)	0.9527		
16		(rf-D+0.5*V²)	0.0065		
17		d1	0.1183		
18		N(d1)	0.5471		
19		N(-d1)	0.4529		
20		d2	-0.1817		
21		N(d2)	0.4279		
22		N(-d2)	0.5721		
23		Probability density function of the Standard Normal Distribution	0.3962		
24					
25		Result	Long Call	Long Put	
26		Option Price	7.67	8.31	
33		Omega	4.7537	-3.6337	
34					

Fig. 2.10 Calculation of the omega

2.10 Assignment 19: Greeks and Other Key Figures at a Glance

Task
Discuss the meaning and application in derivatives practice of the so-called Greeks and the other ratios, Leverage and Omega. Show an overview of the results of the Greeks based on the premises of Assignment 5. Furthermore, show the signs of the Greeks and the further ratios leverage and omega for the four basic strategies (long call, short call, long put, short put) of options (see Chap. 4) and explain this.

Content

Greeks are the sensitivities of the option price to the various value drivers of an option (see assignment 2). They indicate how the option price changes if, for example, the price of the underlying changes (ceteris paribus analysis). The changes are abbreviated with Greek letters. That is why we speak of the "Greeks." Mathematically, the individual derivatives are to be calculated from the Black-Scholes formula. The central Greeks presented here are:

– Delta:	δ or Δ
– Gamma:	γ or Γ
– Vega:	ν
– Theta:	θ
– Rho:	ρ

Depending on which of the four basic option strategies is selected and which of the Greeks are selected, the following sign matrix of the Greeks results (Fig. 2.11):

Important Formulas
The formulas for calculating the Greeks and other ratios were presented in assignments 12 (Delta), 13 (Gamma), 14 (Theta), 15 (Rho), 16 (Vega), 17 (Leverage), and 18 (Omega).

	Delta	Gamma	Theta	Rho	Vega	Lever	Omega
Long-Call	positive	positive	negative	positive	positive	positive	positive
Short-Call	negative	negative	positive	negative	negative	positive	negative
Long-Put	negative	positive	negative	negative	positive	positive	negative
Short-Put	positive	negative	positive	positive	negative	positive	positive

Fig. 2.11 Sign matrix of the Greeks depending on the four basic strategies of options

Black Scholes Merton Model (including the derivation of the Greeks)
Example: Pharma Group

Assumptions	
Price of the underlying	70.00
Strike price	68.00
Dividend	4.85%
Maturity in years	1.00
Implied Volatility (Annual)	30.00%
Risk-free rate	1.00%

Calculations	
Exp(-rf *T)	0.9900
Exp(-DT)	0.9527
(rf-D+0.5*V^2)	0.0065
d1	0.1183
N(d1)	0.5471
N(-d1)	0.4529
d2	-0.1817
N(d2)	0.4279
N(-d2)	0.5721
Probability density function of the Standard Normal Distribution	0.3962

Result	Long Call	Long Put
Option Price	7.67	8.31
Delta	0.5212	-0.4315
Gamma	0.0180	0.0180
Theta	-0.0099	-0.0202
Rho	28.8081	-38.5153
Vega	26.4184	26.4184
Leverage Factor	9.1210	8.4216
Omega	4.7537	-3.6337

Fig. 2.12 Black-Scholes model and Black-Scholes-Merton model

Display in Excel

To calculate the Greeks and other ratios for the Black-Scholes model, simply enter the value "zero" in the "Black-Scholes-Merton" worksheet in the dividend field. This results in the following Greeks and other ratios for both the Black-Scholes model and the Black-Scholes-Merton model (Fig. 2.12).

Further Reading

Ernst, Dietmar; Häcker, Joachim (2016): Financial Modeling, pp. 786–789.
See Excel file Case Study Derivatives Part 1, Excel worksheet 14.
Toolbox, Tasks 117–131.
See video "Black Scholes Model."
Ernst, Dietmar; Häcker, Joachim (2016): Financial Modeling, pp. 789–792.
See Excel file Case Study Derivatives Part 1, Excel worksheet 15.
Toolbox, Tasks 132–141.
See the video "Black Scholes Merton Model," the video "Value drivers and basics of options – Step 3" and the video "Black Scholes Model – Step 3."
Ernst, Dietmar; Häcker, Joachim (2016): Financial Modeling, pp. 792–795.
Hull, John (2015): Options, futures, and other derivatives. S. 503–520.
See Excel file Case Study Derivatives Part 1, Excel worksheet 17–23.
Toolbox, Tasks 151–166.
See video "Greeks."
Ernst, Dietmar; Häcker, Joachim (2016): Financial Modeling, pp. 795–796.
Hull, John (2015): Options, futures, and other derivatives. S. 503–520.
See Excel file Case Study Derivatives Part 1, Excel worksheet 17–23.
Toolbox, Tasks 167–180.
See video "Greeks."
Ernst, Dietmar; Häcker, Joachim (2016): Financial Modeling, pp. 796–797.
Hull, John (2015): Options, futures, and other derivatives. S. 503–520.
See Excel file Case Study Derivatives Part 1, Excel worksheet 17–23.
Toolbox, Tasks 181–192.
See video "Greeks."
Ernst, Dietmar; Häcker, Joachim (2016): Financial Modeling, pp. 797–798.
Hull, John (2015): Options, futures, and other derivatives. S. 503–520.
See Excel file Case Study Derivatives Part 1, Excel worksheet 17–23.
Toolbox, Tasks 193–203.
See video "Greeks."
Ernst, Dietmar; Häcker, Joachim (2016): Financial Modeling, p. 798.
Hull, John (2015): Options, futures, and other derivatives. S. 503–520.
See Excel file Case Study Derivatives Part 1, Excel worksheet 17–23.
Toolbox, Tasks 204–215.
See video "Greeks."
Ernst, Dietmar; Häcker, Joachim (2016): Financial Modeling, pp. 798–799.
Hull, John (2015): Options, futures, and other derivatives. S. 503–520.
See Excel file Case Study Derivatives Part 1, Excel worksheet 17–23.
Toolbox, Tasks 216–221.
See video "Greeks."
Ernst, Dietmar; Häcker, Joachim (2016): Financial Modeling, pp. 799–801.
Hull, John (2015): Options, futures, and other derivatives. S. 503–520.
See Excel file Case Study Derivatives Part 1, Excel worksheet 17–23.
Toolbox, Tasks 222–227.
See video "Greeks."
Ernst, Dietmar; Häcker, Joachim (2016): Financial Modeling, pp. 792–801.
Hull, John (2015): Options, futures, and other derivatives. S. 500–520.
See Excel file Case Study Derivatives Part 1, Excel worksheet 17–23.
Toolbox, Tasks 222–227.
See video "Greeks."

Chapter 3
Fundamentals and Pricing of Futures

3.1 Assignment 20: What Are Futures and Forwards?

Task
Explain what is meant by the terms future and forward. Differentiate between the two terms.

Content
Futures and forwards correspond to a simple purchase transaction in which the traded product is paid for and delivered at a later date. Hedging is achieved by fixing the future price already at the time of trading.

Futures and forwards are forward transactions:

- *Which include the obligation*
- *A certain object (also called the underlying or underlying value)*
- *At a date already set in advance*
- *At a price agreed in advance (future or forward price)*
- *To take over (long) or deliver (short) in a specified quality and quantity (contract)*

In contrast to the options discussed in Chaps. 1 and 2, there is no option right for futures or forwards. The forward transaction must be fulfilled. This is why it is also referred to as an unconditional forward transaction. The period between the conclusion of a future/forward and the delivery date is called the term.

A *future* is a contract whose components are standardized with regard to the payment and delivery terms as well as the tradable size units (contract size). Futures are therefore traded on stock exchanges. Due to this characteristic, they are transferable at any time.

Supplementary Information The online version contains supplementary material available at https://doi.org/10.1007/978-3-031-85822-2_3.

D. Ernst, J. Häcker, *Derivatives*, Springer Texts in Business and Economics,
https://doi.org/10.1007/978-3-031-85822-2_3

A *forward* allows an individual contract solution between two parties. This allows the trading partners with a much greater degree of flexibility with regard to the contract size or delivery date. This type of individualized transaction is called an over-the-counter or OTC transaction for short. The disadvantage of forwards is that they cannot be easily transferred to a third party. While futures are traded daily on futures exchanges and can thus be bought and sold by the contracting parties, forwards cannot be traded and remain as a contract between both contracting parties until they are fulfilled.

The main difference between futures and forwards is that, in contrast to futures, no collateral is provided for forwards. This means that forwards are subject to a settlement risk, as there is no exchange as a central counterparty between the trading partners. The margining of futures offsets the disadvantage of forwards.

In futures or forward hedging, the risk in the current or even the future spot transaction is offset by the financial engineer taking as equal and opposite a position as possible in the forward transaction. For example, a company that has a large quantity of crude oil in stock can hedge against a falling crude oil price by selling the corresponding number of futures contracts on crude oil. These correspond to the company's own crude oil stock. If the price of crude oil falls, the company suffers a loss in value for the stored crude oil. On the other hand, however, this is offset by the short futures position.

3.2 Assignment 21: Which Futures Are Essential for Practice?

Task
Name and describe the most important futures in practice. Use current financial figures for this purpose.

Content
Futures and forwards can be used to hedge various underlyings. A basic distinction is made between financial futures and commodity futures. Figure 3.1 provides an overview of possible financial futures and commodity futures.

The most important futures in practice can be derived from Fig. 3.1. These are:

1. Index futures (indices)
2. Futures on single stocks (shares)
3. Interest rate futures (bonds and money market products)
4. Foreign exchange futures (currencies)
5. Commodity futures (crude oil and gas, metals, food)

Ad 1: Index Futures
In practice, there are many futures on indices. With an index future, it is important to know the underlying precisely:

- Is it a performance or price index?
- How many values are included in the index?

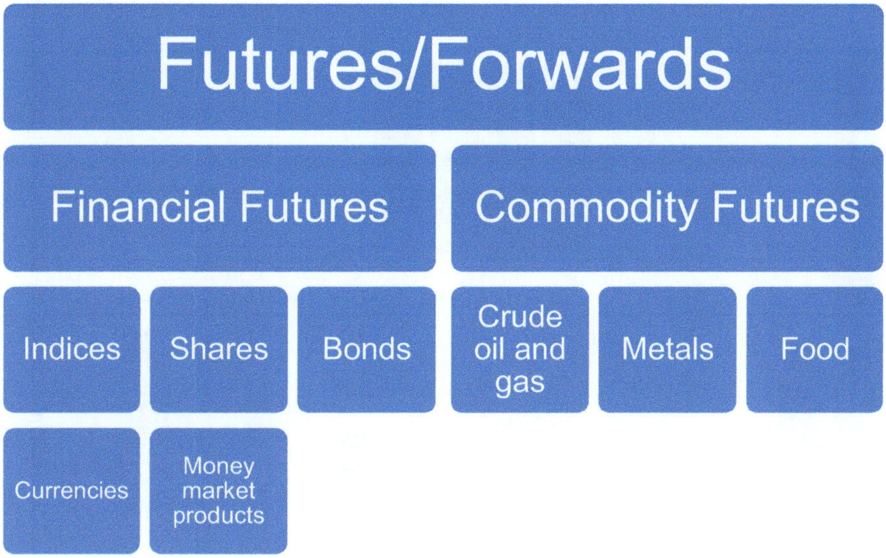

Fig. 3.1 Overview of financial futures and commodity futures

- How is the index calculated?
- What is the index multiplier, and what are the trading hours?

After answering these questions, the actual trading of index futures can begin. Figure 3.2 shows the world's most important financial centers and their respective indices.

The values have been prepared with reference to the reporting date of December 31, 2020. Using the DAX 30 as an example, the statement in the blue boxes is specified in more detail. At that time, the DAX consisted of the 30 largest companies measured by the value of their free float and trading volume. Today, the DAX consists of the 40 largest companies. The market capitalizations of all stocks are added together, resulting in the aggregate market capitalization of all shares included in the index, amounting to EUR 1.25 trillion.

If the financial engineer wants to trade an index future without the respective country risk, he will choose a "world index." In the Global Capital Asset Pricing Model, this corresponds to the so-called market portfolio. In general, the MSCI World is used (see the top line in Fig. 3.2).

The MSCI World Index (shown in blue in Fig. 3.3) includes shares from industrialized countries but not from emerging markets. The industrialized countries are the United States, Canada, Australia, New Zealand, Western Europe (including the United Kingdom and Switzerland), Israel, Japan, Hong Kong, and Singapore. A problem with the MSCI World Index is that the weighting does not correspond to the gross national product generated. For example, the United States accounts for about 2/3 of the total market capitalization of the MSCI World, while the United States "only" has a share of about 22% of the global GDP. This is why there is the MSCI Emerging Markets (shown in orange in Fig. 3.3). It includes the BRIC countries (Brazil, Russia, India, and China) and 23 other countries.

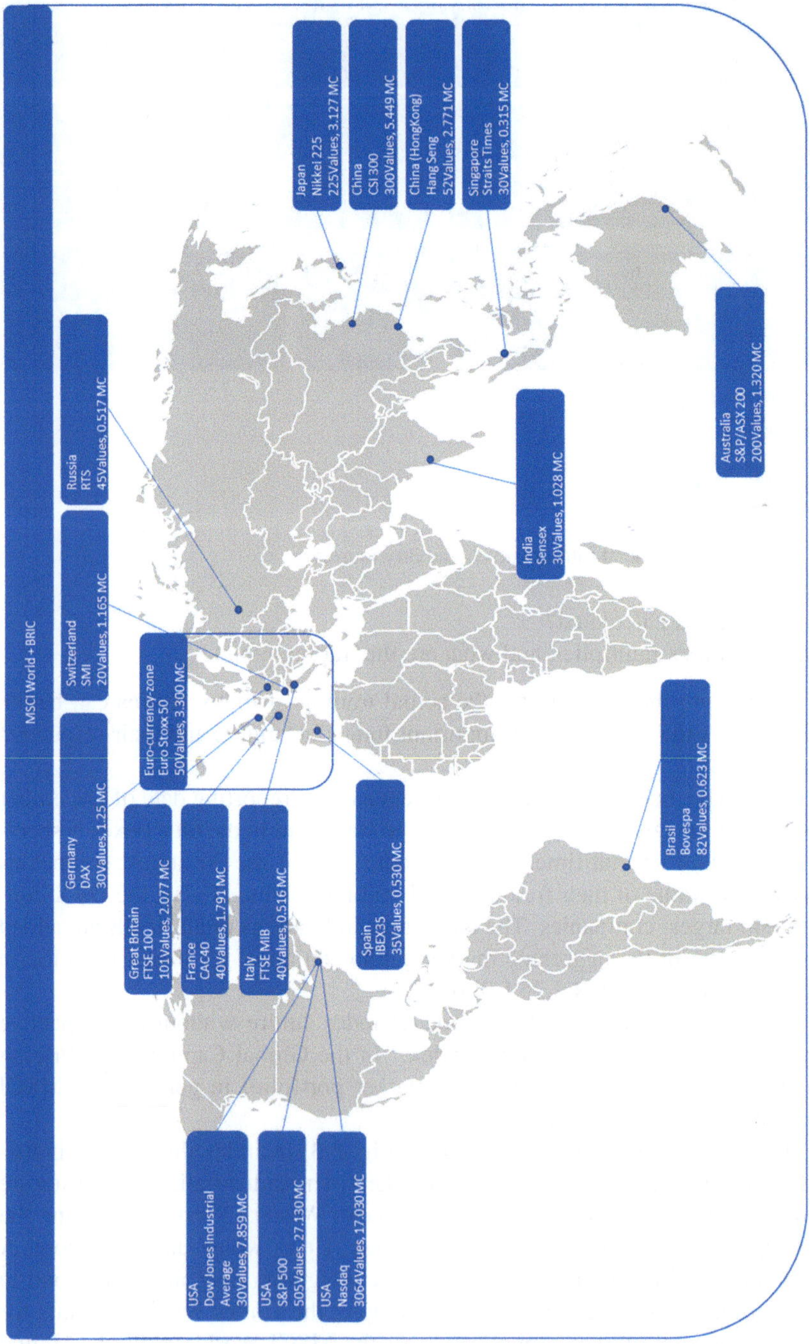

Fig. 3.2 Global financial centers and their indices; total market cap (MC) in trillions of EUR as of 31/12/2020. Source for market cap: Bloomberg. Own illustration

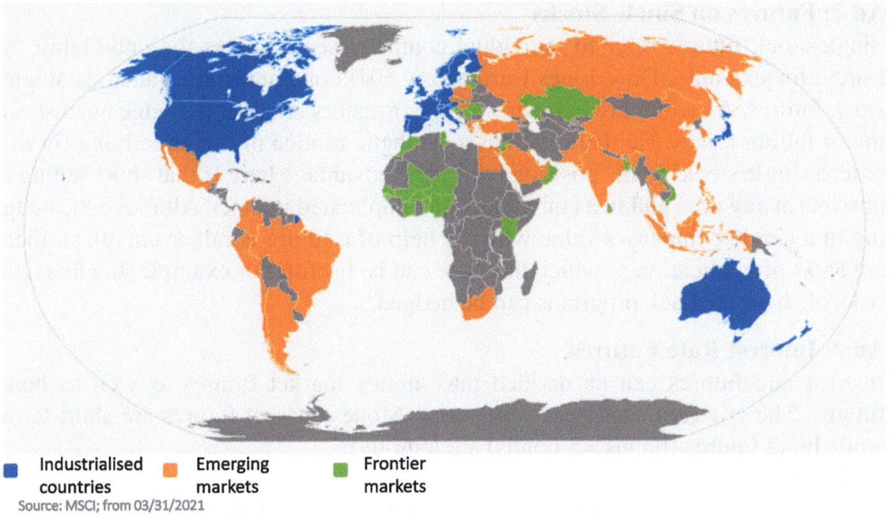

Fig. 3.3 MSCI All Country World Index consisting of MSCI World (blue), MSCI Emerging Markets (orange), and MSCI Frontier Markets (green)

If these countries are added to the MSCI World, the result is the compilation of global financial centers and their indices, as shown in Fig. 3.3. If the MSCI Frontier Markets (shown in green in Fig. 3.3) are added, the result is the MSCI All Country World Index (MSCI ACWI).

The MSCI ACWI contains approximately 85% of the global market capitalization. Approximately 3000 large and medium-sized companies in 23 industrialized countries and 27 emerging markets are represented in the index. The share of emerging markets in the MSCI ACWI amounts to approximately 13%; on the other hand, the share of these countries corresponds to approximately 40% of the global GDP.

Since no index can be delivered, only cash settlement takes place for stock index futures, which is equivalent to a differential payment in cash. In practice, the majority of stock index futures contracts are closed out before the expiration date.

When stock index futures are used for hedging, they are usually used to hedge larger stock portfolios. With a stock index future, a whole basket of individual assets can be hedged with just one transaction. This greatly reduces the time and transaction costs required to hedge all positions individually. However, care must be taken to ensure that the index used corresponds as closely as possible to the composition of the investor's own portfolio. A perfect hedge is hardly possible in practice. In most cases, there is a cross-hedge in which the underlying of the futures only approximates the portfolio to be hedged. One reason for this is that the contract sizes specified by the exchanges may differ from the positions to be hedged. Furthermore, the term of the stock index future may differ from the desired hedging period.

Ad 2: Futures on Single Stocks

Single-stock futures refer to individual company securities as the underlying. At Eurex, for example, Dow Jones Euro Stoxx 50® companies are traded as single-stock futures. As with classic index futures, investors are able to hedge against rising or falling prices. He therefore has a synthetic replica of the underlying (in this case, a single stock) in his position book. The advantage here is that short selling is possible at any time and in a completely uncomplicated manner. Although the hedging of a single company's value with the help of a future is rather untypical, there are areas of application in which their use can be useful. For example, the financial risks of share buyback programs can be hedged.

Ad 3: Interest Rate Futures

Interest rate futures can be divided into money market futures as well as bond futures. The criterion used here is maturity. Money market futures are short-term, while bond futures (bonds => bonds) are long-term.

Money Market Futures

In addition to the above-mentioned characteristics of futures, money market futures have the following special features (Ernst and Häcker 2021):

1. *Underlying:* The underlying is an agreed interest rate for a specific term. In the case of the EURIBOR future on the EUREX, for example, it is the EURIBOR interest rate for the 3-month forward.
2. *Quotation:* The future price reflects the amount of the agreed interest rate. However, the quotation is made in such a way that the corresponding interest rate is subtracted from 100%. If the interest rate is 1.5%, for example, the future price is 98.5% (= 100% − 1.5%). As with bonds, the following also applies to money market futures: the lower the interest rate, the higher the future price. As a result, the buyer of a money market future benefits from falling interest rates, as this increases the price of the money market future. Similarly, the seller of a money market future benefits from rising money market interest rates, since in this case, the prices fall.
3. *Settlement:* As physical delivery is not possible due to the underlying, settlement for interest rate futures in the money market area is always effected by means of cash settlement.

Money market futures are often used to hedge an existing interest rate risk:

Money market futures from the buyer's perspective: From the buyer's perspective, money market futures represent a possibility.

The purchase of money market futures is a strategy to hedge against an expected decline in interest rates. The purchase of money market futures represents a strategy to hedge against an expected decline in interest rates. The expected profit in the futures position resulting from the decline in interest rates exactly offsets the financial disadvantage that arises later in the cash investment due to the lower interest rates. This creates a constellation in which the company is placed in a position as if it would actually make the cash investment on the later date of the cash investment at the investment interest rate agreed today.

Money market futures from the seller's point of view: From the seller's point of view, on the other hand, money market futures represent an opportunity to borrow from the maturity date at an agreed interest rate with an agreed maturity. Selling money market futures is a strategy to hedge against an expected rise in interest rates. The expected profit in the futures position resulting from the interest rate increase exactly offsets the financial disadvantage that arises later when borrowing due to the higher interest rates. This creates a constellation in which the company is placed in a position as if it would actually receive the loan on the later date of borrowing at the loan rate agreed today.

In principle, the interest rate risk of money market futures can be hedged exactly since, in money market futures, the underlying represents an interest rate. However, exact hedging requires that the contract sizes specified by the exchange, the term of the future, and the maturities of the credit match the needs of the hedger.

Bond Futures

Bond futures form the largest futures market. Here, different maturities of the underlying assets are available in order to be able to cover the respective yield expectations. The most heavily traded contracts are the Euro-Bund futures and the 30-year Treasury futures in the United States (futures on U.S. government securities with a 30-year maturity).

Bond futures, such as the Euro-Bund Future (FGBL), are physically delivered. The FGBL is based on a synthetic federal bond with a remaining term of 8.5–10.5 years. The bond future is delivered physically, i.e., effectively. Since the FGBL is a notional federal government bond, a basket of bonds is admitted, which can be used for delivery on the maturity date. However, the bonds do not correspond to the notional 1:1 mapping and must be converted using the price factor, also known as the conversion factor. This creates a basis for comparing different bonds. With the help of the conversion factor, the different coupons and maturities, as well as the standardized contract specifications of the FGBL, can be compensated. It should be noted that a large proportion of futures contracts are not effectively delivered, as they are already closed or rolled over before maturity.

The trading volume and frequency of interest rate futures, such as the Euro-Bund Future and the 30-year Treasury Bond Future (T-Bond), are high. Due to different futures (maturity structures of bonds), investors can profit from changes in the interest rate market, imbalances, and shifts in the yield curve. An investor who expects interest rates to rise (at the long-term end) will sell contracts on the Euro-Bund future. With the help of such an operation, an investor covers both ends of the yield curve and can invest in it by modeling, even if the yield curve shifts.

Ad 4: Foreign Exchange Futures

Foreign exchange futures are also called currency futures or FX futures. At the CME in Chicago, for example, fixed currency pairs such as Euro/USD can be traded. The contract corresponds to a sum of EUR 125,000. An investor can bet on a rise of the Euro against the USD by buying (long), or vice versa, speculate on a fall of the Euro against the USD by selling the futures (short). This relationship also applies to all other currency pairs offered. Due to the fast and liquid trading possibilities, these futures are particularly worthwhile for quick speculations but also for medium- or long-term investment horizons. The possibility of increasing returns can be achieved through short and complementary trades in an active derivatives position book.

Compared with OTC transactions, foreign exchange futures are of very minor importance in foreign exchange trading. Only about 5% of the total market is traded via foreign exchange futures, while 95% of transactions take place directly in interbank trading. The most common currency pairs are the euro to the USD, CHF, and JPY.

Ad 5: Commodity Futures

The history of derivatives is almost 4000 years old (see assignment 1). Until 1973, they were exclusively commodity futures. The history of derivatives is thus a history of commodities. Commodity futures are thus the original reason for the existence of futures exchanges. While in former times the focus was almost exclusively on hedging, in modern times the speculative purpose has also been added. The following commodities play a particularly important role:

Food: In food (soft commodities), futures on wheat, corn, sugar, coffee, and FCOJ (Frozen Concentrated Orange Juice) dominate.

Crude Oil and Gas: Crude Oil and Gas is mostly about Brent, Diesel, Henry Hub Natural Gas, and Light Sweet Crude Oil.

Metals: In the metals sector, the focus is on gold, silver, platinum, palladium, copper, tin, zinc, and iron ore.

3.3 Assignment 22: How Is Futures Pricing Done?

Task

Present in general terms, and with an example, how futures pricing is fundamentally done.

Content

Why do spot and forward rates differ? The only reason why there are differences between spot and forward rates is the different settlement dates of the two transactions. The traded underlying is the same for the hedging transactions. The spot price and forward price diverge because the different settlement dates have specific advantages and disadvantages that are taken into account when pricing the forward. These advantages and disadvantages must be identified and priced. The following advantages and disadvantages can be named.

Financing costs: Compared to spot market transactions, forward purchases do not incur any financing costs until the maturity date. The reason for this is that the underlying does not have to be paid for until the delivery date. Consequently, the future price must be higher than the spot price of the underlying by the amount of the avoided financing costs.

Yields: In the case of a forward transaction, however, there is no possible income from the underlying until the maturity date. These can include, for example, coupon payments in the case of bonds or dividend payments in the case of shares. Since the buyer of a future does not benefit from these payments, he is at a disadvantage compared to buying in the cash market. Consequently, the futures price must be lower than the spot price by the amount of the income accruing up to the maturity date.

Warehousing costs: In the case of commodity futures, warehousing costs are an additional price-determining component. Gold, crude oil, and foodstuffs incur storage costs, which make the forward price more expensive than the spot market price. The commodities must be stored until maturity, which incurs corresponding costs. Consequently, the futures price must be higher than the spot price of the underlying by the amount of the avoided storage costs. The same applies to custody fees for individual shares.

The balance of the three listed advantages and disadvantages is called the cost of carry. The cost of carry (CoC) is the net cost of buying and storing (= "carrying") the underlying until the maturity of a forward transaction.

Important Formulas

$$Cost\ of\ Carry(\in) = financing\ costs - revenues + inventory\ costs$$

Taking into account the cost of carry, we can formulate a general price relationship between the forward price and the current price (spot market price) of the underlying asset:

$$\begin{aligned} F_0 &= X_0 + Cost\ of\ Carry(\in) \\ &= X_0 + financing\ costs - revenues + inventory\ costs \end{aligned}$$

Here, F_0 denotes the current price of the future, and X_0 denotes the current spot price of the underlying. Financing costs and any storage costs increase the forward price, while income from the underlying until maturity reduces the price of the forward compared to X_0.

The formula shows that the price of the forward depends exclusively on the current price X_0 (spot price) plus the cost of carry. If all market participants assume that the price will increase, then the price on the *spot market will* also increase immediately. This means that the expectations are immediately reflected in the prices and not only with a delay in the futures market. Admittedly, the presentation of the pricing of futures has been greatly simplified here. Using interest rate futures (here: money market futures) as an example, we have shown in more detail in our book "Risk Management in Business" how this is done from the background of hedging.

Display in Excel (Fig. 3.4)

Price index future (C16)	= C6+C9-C11
Price interest rate future (C17)	= C6+C9-C11
Price forex future (C18)	= C7+C10-C12
Price Commodity Future (C19)	= C6+C9-C11+C13
Price single value future (C20)	= C8+C9-C11

	A	B	C
1			
2		**Pricing of Futures**	
3		**Example: Pharma Group**	
4			
5		**Assumptions**	
6		Spot price (Index, interest rate, commodity)	100.00
7		Spot price (Curruncy)	1.12
8		Spot price Pharma Group share	48.16
9		Financing costs (Index, interest rate, commodity, single s	2.00
10		Financing costs (Currency)	0.02
11		Income (Index, interest rate, single stocks)	1.50
12		Income (Currency)	0.01
13		Storage costs (Commodity)	0.50
14			
15		**Calculation of the Future price**	
16		Index futures	100.50
17		Interest rate futures	100.50
18		Currency futures	1.13
19		Commodity futures	101.00
20		Futures on single stocks	48.66
21			

Fig. 3.4 Pricing of futures

Further Reading

Ernst, D.; Häcker, J. (2016): Financial Modeling, pp. 810–813.

Bösch, M. (2020): Derivate: Verstehen, anwenden und bewerten, 4th edition, Munich, pp. 177–184.

Hull, J.C. (2014): Risk management: banking, insurance and other financial institutions, 3rd edition, Pearson, pp. 283–295.

Ernst, D.; Häcker, J. (2016): Financial Modeling, pp. 813–816.

Ernst, D.; Häcker, J. (2021): Risikomanagment im Unternehmen, pp. 167-168.

Bösch, M. (2020): Derivate: Verstehen, anwenden und bewerten, 4th edition, Munich, pp. 193–242.

Hull, J.C. (2014): Risk management: banking, insurance and other financial institutions, 3rd edition, Pearson, pp. 283–295.

https://www.justetf.com/de/news/etf/die-msci-index-klassifikationen-und-wie-sie-die-welt-einteilen.html

Ernst, D.; Häcker, J. (2016): Financial Modeling, pp. 816–825.

Ernst, D.; Häcker, J. (2021): Risk management in business – step by step, pp. 165–177.

See Excel file Case Study Derivatives Part 1, Excel worksheet 29.

Part II
Option Strategies

1. Black Scholes Merton Model including the derivation of the Greeks (Strike = 68)

Assumptions	
Price of the underlying	70.00
Strike price	68.00
Dividend	4.85%
Maturity in years	1.00
Implied Volatility (Annual)	30.00%
Risk-free rate	1.00%

Calculation	
Exp(-rf * T)	0.9900
Exp(-DT)	0.9527
(rf-D+0.5*V²)	0.0065
d1	0.1183
N(d1)	0.5471
N(-d1)	0.4529
d2	-0.1817
N(d2)	0.4279
N(-d2)	0.5721
Distribution function of the Standard Normal Distribution	0.3962

Result	Long Call	Long Put
Option Price	7.67	8.31
Delta	0.5212	-0.4315
Gamma	0.0180	0.0180
Theta	-0.0098	-0.0202
Rho	28.8081	-38.5153
Vega	26.4184	26.4184
Leverage Factor	9.1210	8.4216
Omega	4.7537	-3.5337

2. Black Scholes Merton Model including the derivation of the Greeks (Strike = 70)

Assumptions	
Price of the underlying	70.00
Strike price	70.00
Dividend	#####
Maturity in years	1.00
Implied Volatility (Annual)	######
Risk-free rate	#####

Calculation	
Exp(-rf * T)	0.9900
Exp(-DT)	0.9527
(rf-D+0.5*V²)	0.0065
d1	0.0217
N(d1)	0.5086
N(-d1)	0.4914
d2	-0.2783
N(d2)	0.3904
N(-d2)	0.6096
Distribution function of the Standard Normal Distribution	0.3988

Result	Long Call	Long Put
Option Price	6.86	9.48
Delta	0.4846	-0.4681
Gamma	0.0181	0.0181
Theta	-0.0105	-0.0208
Rho	27.0546	-42.2489
Vega	26.5976	26.5976
Leverage Factor	10.1969	7.3822
Omega	4.9411	-3.4596

3. Black Scholes Merton Model including the derivation of the Greeks (Strike = 72)

Assumptions	
Price of the underlying	70.00
Strike price	72.00
Dividend	#####
Maturity in years	1.00
Implied Volatility (Annual)	######
Risk-free rate	#####

Calculation	
Exp(-rf * T)	0.9900
Exp(-DT)	0.9527
(rf-D+0.5*V²)	0.0065
d1	(0.0722)
N(d1)	0.4712
N(-d1)	0.5288
d2	-0.3722
N(d2)	0.3549
N(-d2)	0.6451
Distribution function of the Standard Normal Distribution	0.3979

Result	Long Call	Long Put
Option Price	6.13	10.72
Delta	0.4489	-0.5038
Gamma	0.0181	0.0181
Theta	-0.0108	-0.0209
Rho	25.2956	-45.9880
Vega	26.5346	26.5346
Leverage Factor	11.4242	6.5269
Omega	5.1283	-3.2580

Fig. 4.2 Calculation of the respective option premiums depending on the strike price

4.3 Assignment 3: Short Call

Task

Calculate what profit the investor of a short call achieves with a call price of EUR 7.67 and a strike in the amount of EUR 68.00. The price of the call is again based on the Black-Scholes-Merton model. At what point would the call be exercised? Explain the maximum profit and loss. Possible developments of the Pharma Group share are 0.00 EUR to 140.00 EUR in 5 EUR steps. Plot the result graphically.

Content

At the strike of EUR 68.00, the buyer of the call would let the option expire if the price of the Pharma Group share is below EUR 68. The financial engineer would thus receive the premium of EUR 7.67 as his maximum possible profit.

If the price of the underlying asset rises above EUR 68.00, the buyer will exercise the option, and the financial engineer's profit will thus be reduced. The break-even point is 75.67 EUR. If the price of the underlying asset rises above this, the financial engineer will incur a loss. Under the premise of a "physical delivery" (not "cash settlement"), the holder of a short call is obliged to deliver the Pharma Group share to the holder of the long position at the strike. On the other hand, the Pharma Group share price can theoretically rise to infinity. Thus, the risk of loss for the Financial Engineer is also infinite.

In the case of short calls, a distinction can be made between naked call writing and covered call writing. In the case of naked call writing, the seller of the call does not own the underlying. Therefore, if the call is exercised by the buyer and the seller of the short call is thus obliged to deliver it, he must first cover himself with it on the open market.

In the case of covered call writing, on the other hand, the seller of the short call owns the underlying and can deliver it from his own holdings in the event of exercise by the buyer (see assignments 7 and 8).

Long Call: Profit and loss

Assumptions	
Price of the Call at closing	7.67
Strike price	68.00

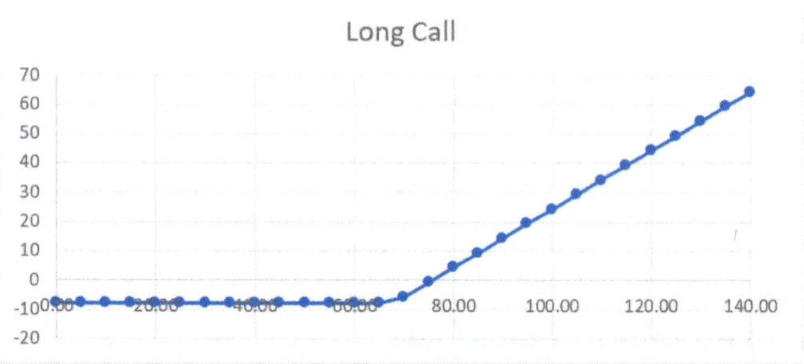

Price of the underlying	Profit	Will the option be exercised?
0.00	(7.67)	no
5.00	(7.67)	no
10.00	(7.67)	no
15.00	(7.67)	no
20.00	(7.67)	no
25.00	(7.67)	no
30.00	(7.67)	no
35.00	(7.67)	no
40.00	(7.67)	no
45.00	(7.67)	no
50.00	(7.67)	no
55.00	(7.67)	no
60.00	(7.67)	no
65.00	(7.67)	no
70.00	(5.67)	yes
75.00	(0.67)	yes
80.00	4.33	yes
85.00	9.33	yes
90.00	14.33	yes
95.00	19.33	yes
100.00	24.33	yes
105.00	29.33	yes
110.00	34.33	yes
115.00	39.33	yes
120.00	44.33	yes
125.00	49.33	yes
130.00	54.33	yes
135.00	59.33	yes
140.00	64.33	yes

Fig. 4.3 Calculation of long call

Important Formulas

$$\text{Exercise decision}(\text{yes}) = \text{price of the underlying} - \text{strike} > \text{zero}$$

$$\text{Profit short call} = \text{strike} - \text{price of the underlying} + \text{price of the call}$$

Display in Excel
You can enter the following formulas in the two cells C20 and D20:

What is the profit? => C20 =IF(B20>$C$6;$C$6- $B20+$C$5;$C$5)
Will the option be exercised? => D20 =IF(B20>$C$6;"yes";"no")

Following this, you can drag the cell C20 to C48 and the cell D20 to D48 with the mouse. You will then get the following picture as shown in Fig. 4.4.

4.4 Assignment 4: Long Put

Task
Calculate the profit/loss of the investor of a long put at a put price of EUR 10.72 and a strike of EUR 72. The price of the put is again based on the Black-Scholes-Merton model. At what point would the investor exercise the put? Possible developments of the Pharma Group share range from 0.00 EUR to 140.00 EUR in 5 EUR steps. Plot the result graphically.

Content
At the current Pharma Group share price of EUR 70 and the strike of EUR 72.00, the buyer of the put option would exercise the option. The break-even point is reached at a share price of EUR 61.28 (72-10.72). The more the share price falls, the more the profit of the buyer of the put now increases. The profit is limited to 61.28 EUR (72-0-10.72) because the share price of the Pharma Group can be at least zero and cannot become negative.

Important Formulas

$$\text{Exercise decision}(\text{yes}) = \text{price of the underlying} - \text{strike} < \text{zero}$$

$$\text{Profit long put} = \text{strike} - \text{price of the underlying} - \text{price of the put}$$

$$\text{Maximum loss long put} = -\text{price of the put}$$

Short Call: Profit and loss

Assumptions	
Price of the Call at closing	7.67
Strike price	68.00

Price of the underlying	Profit	Will the option be exercised?
0.00	7.67	no
5.00	7.67	no
10.00	7.67	no
15.00	7.67	no
20.00	7.67	no
25.00	7.67	no
30.00	7.67	no
35.00	7.67	no
40.00	7.67	no
45.00	7.67	no
50.00	7.67	no
55.00	7.67	no
60.00	7.67	no
65.00	7.67	no
70.00	5.67	yes
75.00	0.67	yes
80.00	(4.33)	yes
85.00	(9.33)	yes
90.00	(14.33)	yes
95.00	(19.33)	yes
100.00	(24.33)	yes
105.00	(29.33)	yes
110.00	(34.33)	yes
115.00	(39.33)	yes
120.00	(44.33)	yes
125.00	(49.33)	yes
130.00	(54.33)	yes
135.00	(59.33)	yes
140.00	(64.33)	yes

Fig. 4.4 Short call calculation

Display in Excel
You can enter the following formulas in the two cells C20 and D20:

What is the profit? => C20 =IF (B20<$C$6;$C$6-$B20-$C$5;-$C$5)
Is the option exercised? => D20 =IF (B20<$C$6; "yes"; "no")

Following this, you can drag the cell C20 to C48 and the cell D20 to D48 with the mouse. You will then get the following picture as shown in Fig. 4.5.

4.5 Assignment 5: Short Put

Task
Calculate what profit the investor of a short put achieves with a put price of EUR 10.72 and a strike in the amount of EUR 72.00. The price of the put is again based on the Black-Scholes-Merton model. At what point would the put be exercised? Explain the maximum profit and loss. Discuss what should be considered in practice for the short put position. Possible developments of the Pharma Group share range from 0.00 EUR to 140.00 EUR in 5 EUR steps. Present the result graphically.

Content
At the strike of EUR 72.00, the buyer of the put would let the option expire if the price of the Pharma Group share is above EUR 72. The financial engineer would thus receive the premium of EUR 10.72 as his maximum possible profit.

If the price of the underlying falls below EUR 72.00, the buyer will exercise the option, and the financial engineer's profit will thus be reduced. The break-even point is 61.28 EUR (72-10.72). If the price of the underlying falls below 61.28 EUR, the financial engineer will incur a loss.

It should be noted that, in the event of an exercise by the purchaser, the Financial Engineer is obliged to purchase the underlying from the purchaser (in the case of Physical Delivery) and thus to take the underlying into its own portfolio.

A short put should only be chosen by the financial engineer if he is also prepared to take the positions into his own portfolio. This allows for an active build-up of portfolio positions, including a premium income. The financial engineer thus has an entry price for the underlying position reduced by the premium. However, when entering into the option position, he bears the risk of having to acquire the Pharma Group share. The longer the option term, the greater the risk that the assessment of the underlying will change.

Important Formulas

$$\text{Exercise decision}\left(\text{yes}\right) = \text{price of the underlying} - \text{strike} < \text{zero}$$

$$\text{Profit short put} = -\text{strike} + \text{price of the underlying} + \text{price of the put}$$

Long Put: Profit and loss

Assumptions	
Price of the Put at closing	10.72
Strike price	72.00

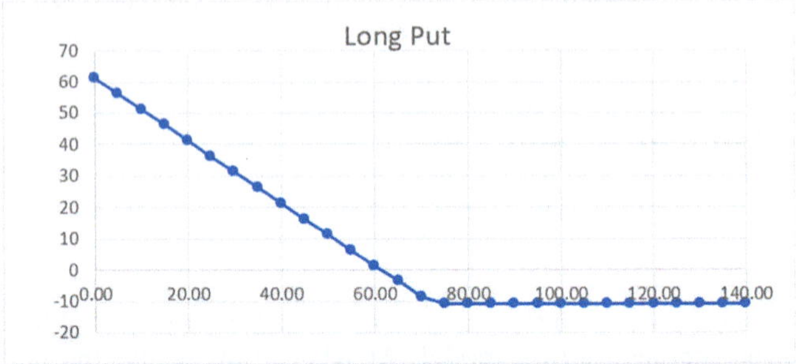

Price of the underlying	Profit	Will the option be exercised?
0.00	61.28	yes
5.00	56.28	yes
10.00	51.28	yes
15.00	46.28	yes
20.00	41.28	yes
25.00	36.28	yes
30.00	31.28	yes
35.00	26.28	yes
40.00	21.28	yes
45.00	16.28	yes
50.00	11.28	yes
55.00	6.28	yes
60.00	1.28	yes
65.00	(3.72)	yes
70.00	(8.72)	yes
75.00	(10.72)	no
80.00	(10.72)	no
85.00	(10.72)	no
90.00	(10.72)	no
95.00	(10.72)	no
100.00	(10.72)	no
105.00	(10.72)	no
110.00	(10.72)	no
115.00	(10.72)	no
120.00	(10.72)	no
125.00	(10.72)	no
130.00	(10.72)	no
135.00	(10.72)	no
140.00	(10.72)	no

Fig. 4.5 Calculation of long put

Display in Excel
You can enter the following formulas in the two cells C20 and D20:

What is the profit? => C20 =IF (B20<$C$6;$B20-$C$6+$C$5;$C$5)
Is the option exercised? => D20 =IF (B20<$C$6; "yes"; "no")

Following this, you can drag the cell C20 to C48 and the cell D20 to D48 with the mouse. You will then get the following picture as shown in Fig. 4.6.

4.6 Assignment 6: Advanced Strategies

Task
How can advanced strategies be formed with options, what are the basic distinguishing features of the strategies, and what are the Advanced Strategies?

Content
How can advanced strategies with options be formed? Basically, any strategy with options is a certain combination of the four basic strategies or buying or selling the underlying asset itself.

What are the basic distinguishing features of the strategies? Depending on his expectations, an investor can decide which developments his strategy should target. On the one hand, he has the possibility to target the development of the underlying asset (whether it rises, falls, or remains constant) itself. Strategies that profit from a rising underlying are called bullish strategies; vice versa, they are called bearish strategies. Another possibility is the so-called neutral strategies. Here, the volatility of the underlying is bet on.

What are the Advanced Strategies? The Advanced Options Strategies extend the basic strategies. With the Advanced Options Strategies, you can distinguish:

1. Bullish advanced options strategies
2. Bearish advanced options strategies and
3. Neutral advanced options strategies

The latter again divide into:

3a) Neutral bearish volatility strategies, as well as
3b) Neutral bullish volatility strategies

The further breakdown is as follows:
Ad 1: Bullish

1. Covered Calls OTM/ITM
2. Protective Put
3. Collar Strategy
4. Bull Call Spread
5. Bull Put Spread
6. Call Backspread

Short Put: Profit and loss

Assumptions	
Price of the Put at closing	10.72
Strike price	72.00

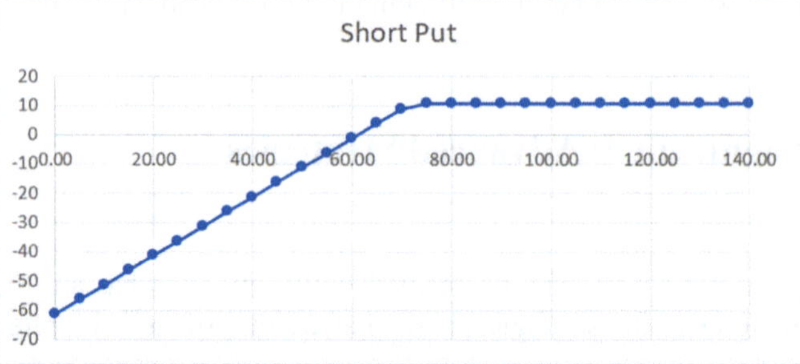

Price of the underlying	Profit	Will the option be exercised?
0.00	(61.28)	yes
5.00	(56.28)	yes
10.00	(51.28)	yes
15.00	(46.28)	yes
20.00	(41.28)	yes
25.00	(36.28)	yes
30.00	(31.28)	yes
35.00	(26.28)	yes
40.00	(21.28)	yes
45.00	(16.28)	yes
50.00	(11.28)	yes
55.00	(6.28)	yes
60.00	(1.28)	yes
65.00	3.72	yes
70.00	8.72	yes
75.00	10.72	no
80.00	10.72	no
85.00	10.72	no
90.00	10.72	no
95.00	10.72	no
100.00	10.72	no
105.00	10.72	no
110.00	10.72	no
115.00	10.72	no
120.00	10.72	no
125.00	10.72	no
130.00	10.72	no
135.00	10.72	no
140.00	10.72	no

Fig. 4.6 Calculation of short put

Ad 2: Bearish

1. Covered Put
2. Put Backspread
3. Bear Put Spread
4. Bear Call Spread
5. Protective Call

Ad 3a: Neutral Bearish

1. Condor
2. Long Call Butterfly
3. Long Put Butterfly
4. Long Call Ladder
5. Long Put Ladder
6. Short Strangle
7. Short Straddle
8. Short Guts

Ad 3b: Neutral Bullish

1. Short Condor
2. Short Call Butterfly
3. Short Put Butterfly
4. Short Call Ladder
5. Short Put Ladder
6. Long Strangle
7. Long Straddle
8. Strip
9. Strap
10. Long Guts

Point 4 is actually arbitrage strategies. However, arbitrage strategies will not be discussed further in the following. The background to this is that arbitrage opportunities are becoming increasingly difficult to realize due to increasingly computerized trading.

The option strategies presented here do not represent the entire possible range of option strategies. From a theoretical point of view, there are very many possibilities to combine the four basic strategies or the purchase/sale of the underlying itself. The strategies presented here represent only a small selection of the theoretically possible combinations. We have based our selection of option strategies on what we consider to be the most important option strategies from a practical point of view. The scientific literature on option strategies is extensive. From a practical point of view, we would like to highlight the "Option Strategy Finder" at www.theoptionsguide.com/option-trading-strategies.aspx. This was a good help to us for the following description of Advanced Strategies. What we have not been able to find so far, neither in scientific publications nor in practical publications, was an overview of option strategies, which, on the basis of a case, holistically and at the same time

Table 4.1 The calls and puts underlying the advanced strategies and their prices

Art	Strike	Price (according to Black-Scholes-Merton)
Put	72.00	10.72
Put	70.00	9.48
Put	**68.00**	**8.31**
Call	**68.00**	**7.67**
Call	70.00	6.86
Call	72.00	6.13

with concrete figures in Excel, shows how the individual aspects are connected and how, with the naming of concrete figures, the individual option strategies can be controlled. This literature gap is to be closed with the following analysis—in particular, the Chap. 6.

For the strategies to be calculated below, the previous assumptions are supplemented as shown in Table 4.1.

Price of the underlying	EUR 70.00
Dividends	4.85%
Term of the option in years	1.00
Implied volatility	30.00%
Risk-free interest rate (rf)	1.00%

Table 4.1 (for more detailed information, see also Fig. 4.2) shows that the case from Part I, highlighted here in bold (strike = EUR 68), is joined by two further cases (strike = EUR 70 and EUR 72).

The basic premise of the following analysis is that the expiration date of the option is assumed to be the observation period. In concrete terms, this means that a time value of zero is assumed. Accordingly, the value of an option is determined exclusively by its intrinsic value.

4.7 Assignment 7: Bullish–Covered Call OTM

Task

(a) How is a Covered Call OTM formed, and under what name is this strategy still known?
(b) What is the loss or gain potential of this strategy?
(c) Calculate the result of a covered calls OTM strategy when buying a Pharma Group share at 70.00 EUR and selling an OTM call. Possible developments of the Pharma Group share range from 0.00 EUR to 140.00 EUR in steps of five. Show the result graphically.
(d) Show which elements make up the Covered Call OTM and how the individual values of the Covered Call OTM can be determined for the respective different share prices of the Pharma Group share.

Content

The strategy Covered Calls, formed with OTM options, is also called Synthetic Short Put because the strategy consists not only of options but also, in its construction, of the underlying asset itself. The advantage of a synthetic financial product is usually that it is cheaper than its pure counterpart. The covered call OTM is formed by the investor himself owning the underlying asset and selling OTM calls at the same time. In the two figures below, it is clear how the Covered Call OTM (blue line) is composed of the Pharma Group share (green line) and the short call on the Pharma Group (orange line). The profit potential of this strategy is limited, whereas the loss potential is unlimited. Arithmetically, the limit of the loss potential is at a Pharma Group share price of zero.

Through the premiums, the investor secures a positive cash flow. In case of an exercise by the counter-position (long call), the risk is limited, as the investor already owns the pieces. With the help of the premium income, which he receives as extraordinary income, the investor is protected against slight price declines.

Achieving a second dividend with short calls: Covered calls OTM offer an investor the opportunity to capitalize on his portfolio holdings, i.e., to generate active returns in addition to dividends and share price appreciation. This can be achieved by the financial engineer writing calls on the inventories spread over the year. If these are called, he can deliver from his own stock. If the stocks are not called, he has secured a kind of second dividend. It is important that the number of calls written does not exceed the number of shares held at any time; otherwise, the short call position is uncovered.

To understand option strategies, the financial engineer should always take a closer look at the strike. The covered call OTM strategy reaches the maximum profit at the strike of the short call: 72 EUR (due to the five steps chosen in the chart below, this is only approximately visible).

- If the share rises above the strike, the counterparty will exercise the call, and the financial engineer will no longer be able to profit from further gains in the share price.
- If the share price falls below EUR 63.87 (70-6.13), the loss in value of the share can no longer be offset by the call premium received.

Thus, this option strategy makes sense if the Financial Engineer wants to make unmoved stocks higher-yielding through premium income.

Important Formulas

$$\text{Maximum profit} = \text{premium received} - \text{purchase price of the underlying} \\ + \text{strike price of the short call}$$

$$\text{Break} - \text{even} = \text{purchase price of the underlying} - \text{premium received}$$

$$\text{Loss} = \text{price of the underlying} - \text{purchase price of the underlying} + \text{premium received}$$

Covered Call OTM

Assumptions	
# underlying	1.00
Price of the underlying	70.00
# calls	1.00
Call price	6.13
Strike price	72.00

Price of the underlying	Profit/loss
0.00	(63.87)
5.00	(58.87)
10.00	(53.87)
15.00	(48.87)
20.00	(43.87)
25.00	(38.87)
30.00	(33.87)
35.00	(28.87)
40.00	(23.87)
45.00	(18.87)
50.00	(13.87)
55.00	(8.87)
60.00	(3.87)
65.00	1.13
70.00	6.13
75.00	8.13
80.00	8.13
85.00	8.13
90.00	8.13
95.00	8.13
100.00	8.13
105.00	8.13
110.00	8.13
115.00	8.13
120.00	8.13
125.00	8.13
130.00	8.13
135.00	8.13
140.00	8.13

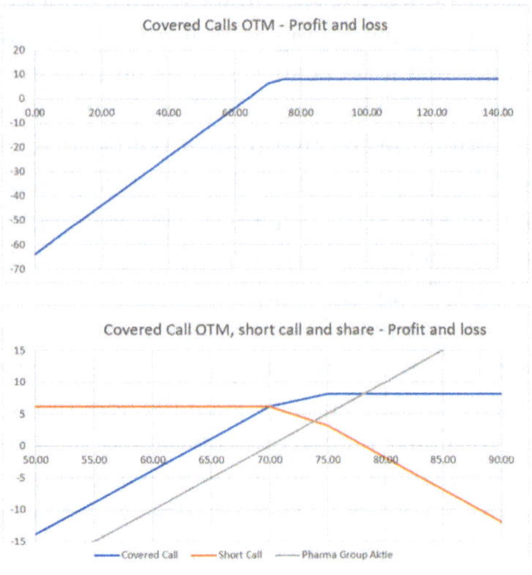

Fig. 4.7 Calculation of covered call OTM

Display in Excel

You can enter the following formula in cell C20:

=SUM (((B20-$C$7)*$C$6);(IF(B20>$C$10;$C$10-B20+$C$9;$C$9)*$C$8))

Following this, you can drag the cell C20 to C48 with the mouse. You will then get the following picture as shown in Fig. 4.7.

The formulas for P20, T20, W20, and V20 are as follows:

P20: =IF(O20>$P$10;$P$10-$O20+$P$9;$P$9)
T20 =S20-$T$7
W20 =T20+P20
V20 =W20-C20

After that, you can drag the respective cell with the mouse to P48, T48, W48, as well as V48. You will then get the following picture as shown in Fig. 4.8.

Short Call

Assumptions	
Price of the underlying	70.00
Call price	6.13
Strike price	72.00

Pharma Group share

Assumptions	
Price of the underlying	70.00
Strike price	72.00

Covered Call OTM
versus
Long share +Short Call

Price of the underlying	Profit/loss	Price of the underlying	Profit/loss	Difference	Profit/loss
0.00	6.13	0.00	(70.00)	-	(63.87)
5.00	6.13	5.00	(65.00)	-	(58.87)
10.00	6.13	10.00	(60.00)	-	(53.87)
15.00	6.13	15.00	(55.00)	-	(48.87)
20.00	6.13	20.00	(50.00)	-	(43.87)
25.00	6.13	25.00	(45.00)	-	(38.87)
30.00	6.13	30.00	(40.00)	-	(33.87)
35.00	6.13	35.00	(35.00)	-	(28.87)
40.00	6.13	40.00	(30.00)	-	(23.87)
45.00	6.13	45.00	(25.00)	-	(18.87)
50.00	6.13	50.00	(20.00)	-	(13.87)
55.00	6.13	55.00	(15.00)	-	(8.87)
60.00	6.13	60.00	(10.00)	-	(3.87)
65.00	6.13	65.00	(5.00)	-	1.13
70.00	6.13	70.00	-	-	6.13
75.00	3.13	75.00	5.00	-	8.13
80.00	(1.87)	80.00	10.00	-	8.13
85.00	(6.87)	85.00	15.00	-	8.13
90.00	(11.87)	90.00	20.00	-	8.13
95.00	(16.87)	95.00	25.00	-	8.13
100.00	(21.87)	100.00	30.00	-	8.13
105.00	(26.87)	105.00	35.00	-	8.13
110.00	(31.87)	110.00	40.00	-	8.13
115.00	(36.87)	115.00	45.00	-	8.13
120.00	(41.87)	120.00	50.00	-	8.13
125.00	(46.87)	125.00	55.00	-	8.13
130.00	(51.87)	130.00	60.00	-	8.13
135.00	(56.87)	135.00	65.00	-	8.13
140.00	(61.87)	140.00	70.00	-	8.13

Fig. 4.8 Representation of the congruence of covered call OTM and long Pharma Group share and short call

4.8 Assignment 8: Bullish–Covered Calls ITM

Task

Explain the differences between the Covered Calls OTM and Covered Calls ITM strategies.

Content

The strategy Covered Calls ITM works analogously to the strategy Covered Calls OTM. The difference is that now, instead of out-of-the-money calls, in-the-money calls are sold by the investor of this strategy. For example, the strike is now reduced from 72 EUR to 64 EUR, and all other input parameters remain unchanged. According to the Black-Scholes-Merton model, this results in a call price of 9.53 EUR. If the share price of the Pharma Group now falls to zero, the ITM call results in a loss of 60.47 EUR (for the OTM call, it was 63.87 EUR). Thus, this strategy still has a loss risk that should not be underestimated, but a lower loss profile compared to the covered calls OTM.

The premium collected from the sale of the ITM call now amounts to EUR 9.53 and is thus higher than for an OTM call listed in the previous assignment (EUR 6.13). The higher premium can thus better compensate for the loss from the security in the event of a price decline.

On the other hand, however, a Pharma Group share price above the strike (now already from EUR 64 and not EUR 72) results in a low profit. The maximum profit that can now be achieved is 3.53 EUR instead of 8.13 EUR.

Important Formulas

$$\text{Maximum profit} = \text{premium received} - \text{purchase price of the underlying} \\ + \text{strike price of the short call}$$

$$\text{Break} - \text{even} = \text{purchase price of the underlying} - \text{premium received}$$

$$\text{Loss} = \text{price of the underlying} - \text{purchase price of the underlying} + \text{premium received}$$

4.9 Assignment 9: Bullish–Protective Put

Task

(a) How is a protective put formed, and what is its typical purpose?
(b) What is the loss or gain potential of this strategy?
(c) Calculate the result of a protective put strategy when buying an ATM put and an underlying. Possible developments of the Pharma Group share are 0.00 EUR to 140.00 EUR in steps of five. Plot the result graphically.
(d) Show which elements make up the bullish Protective Put and how the individual values of the Protective Put can be determined for the respective different share prices of the Pharma Group share.

Content

The Protective Put is a form of Synthetic Long Call. This strategy consists of being long both in the underlying itself and in a put on the underlying. This strategy can therefore also be classified as a classic hedging strategy. The financial engineer continues to assume a rising underlying, but wants to secure his previous price gain or generally hedge against upcoming uncertainties. The profit potential is unlimited, whereas the loss potential is limited.

The profit profile of the Protective Put corresponds, as shown in the figure below, to that of a Long Call plus the Term $X \times e^{-r_f \times T} + D$.

Thus, the put-call parity shown in Part I applies:

$$p + S_0 = c + X \times e^{-r_f \times T} + D$$

This relationship shall be illustrated by the case in which the share price of Pharma Group has increased from EUR 70 to EUR 100. In this case, the value of the protective put (i.e., the left side of the put-call parity) is 20.52 EUR. This is composed of the value of the put in the amount of −9.48 EUR and the value of the share in the amount of 30 EUR. The value of the call (first term on the right-hand side of the put-call parity) is 22.33 EUR. The difference between 20.52 EUR and 22.33 EUR is represented by the term $X \times e^{-r_f \times T} + D$. This is the present value of the fixed-interest investment in the amount of the strike price (X) of the call option, which in turn was discounted continuously. Furthermore, the dividend yield was added to this.

Important Formulas

$$\text{Profit} = \text{price of the underlying} - \text{purchase price of the underlying} - \text{premium paid}$$

$$\text{Break} - \text{even} = \text{purchase price of the underlying} + \text{premium paid}$$

$$\text{Maximum loss} = -\text{premium paid} + \text{purchase price of the underlying} - \text{strike price put}$$

Display in Excel

You can enter the following formula in cell C20:

=SUM (((B20-$C$7)*$C$6);IF(B20<$C$10;$C$10-B20-$C$9;-$C$9)*$C$8)

Following this, you can drag the cell C20 to C48 with the mouse. You will then get the following picture as shown in Fig. 4.9.

The formulas for P20, S20, U20, V20, Y20, AA20, and AB20 are as follows:

P20: =IF(O20<$P$10;$P$10-$O20-$P$9;-$P$9)
S20: =R20-$S$7
U20: =V20-C20
V20: =S20+P20
Y20: =IF(X20>$Y$10;X20-$Y$10-$Y$9;-$Y$9)
AA20: =P20+S20
AB20: =Y20

Subsequently, you can drag the respective cell with the mouse up to row 48. You will then get the following picture as shown in Fig. 4.10.

Protective Put

Assumptions	
# underlying	1.00
Price of the underlying	70.00
# Long Puts	1.00
Long Put price	9.48
Strike price of Long Put	70.00

Price of the underlying	Profit/loss
0.00	(9.48)
5.00	(9.48)
10.00	(9.48)
15.00	(9.48)
20.00	(9.48)
25.00	(9.48)
30.00	(9.48)
35.00	(9.48)
40.00	(9.48)
45.00	(9.48)
50.00	(9.48)
55.00	(9.48)
60.00	(9.48)
65.00	(9.48)
70.00	(9.48)
75.00	(4.48)
80.00	0.52
85.00	5.52
90.00	10.52
95.00	15.52
100.00	20.52
105.00	25.52
110.00	30.52
115.00	35.52
120.00	40.52
125.00	45.52
130.00	50.52
135.00	55.52
140.00	60.52

Received/paid premium	(9.48)

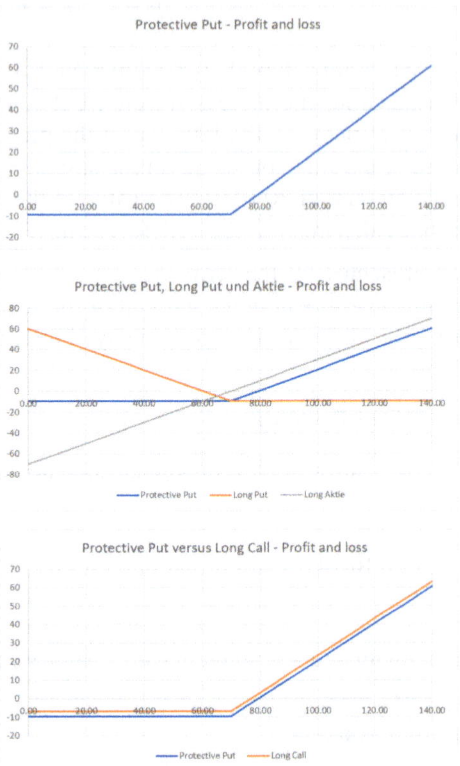

Fig. 4.9 Calculation of Protective Put

4.10 Assignment 10: Bullish–Collar Strategy

Task

(a) How is a collar formed, and what is the loss or profit potential of this strategy?
(b) Calculate the result of a collar strategy when buying an OTM put, an underlying, and selling an OTM call. Possible developments of the Pharma Group share range from 0.00 EUR to 140.00 EUR in steps of five. Plot the result graphically.
(c) Show which elements make up the collar and how the individual values of the collar can be determined for the respective different share prices of the Pharma Group share.

Content

The Collar strategy is suitable for an investor who follows the covered call strategy but at the same time wants to hedge against price losses of the underlying. The

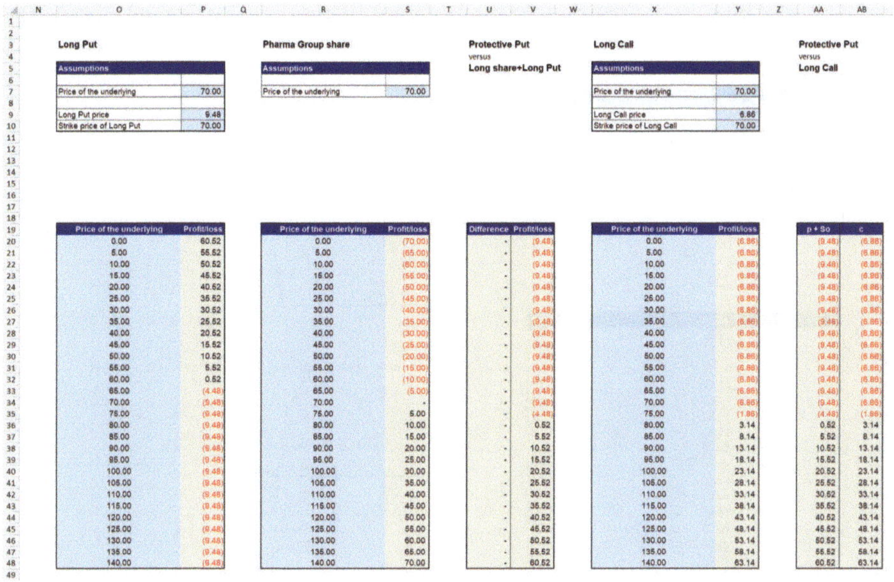

Long Put

Assumptions	
Price of the underlying	70.00
Long Put price	9.48
Strike price of Long Put	70.00

Pharma Group share

Assumptions	
Price of the underlying	70.00

Protective Put versus **Long share+Long Put**

Long Call

Assumptions	
Price of the underlying	70.00
Long Call price	6.86
Strike price of Long Call	70.00

Protective Put versus **Long Call**

Price of the underlying	Profit/loss	Price of the underlying	Profit/loss	Difference	Profit/loss	Price of the underlying	Profit/loss	p + So	c
0.00	60.52	0.00	(70.00)	-	(9.48)	0.00	(6.86)	(9.48)	(6.86)
5.00	55.52	5.00	(65.00)	-	(9.48)	5.00	(6.86)	(9.48)	(6.86)
10.00	50.52	10.00	(60.00)	-	(9.48)	10.00	(6.86)	(9.48)	(6.86)
15.00	45.52	15.00	(55.00)	-	(9.48)	15.00	(6.86)	(9.48)	(6.86)
20.00	40.52	20.00	(50.00)	-	(9.48)	20.00	(6.86)	(9.48)	(6.86)
25.00	35.52	25.00	(45.00)	-	(9.48)	25.00	(6.86)	(9.48)	(6.86)
30.00	30.52	30.00	(40.00)	-	(9.48)	30.00	(6.86)	(9.48)	(6.86)
35.00	25.52	35.00	(35.00)	-	(9.48)	35.00	(6.86)	(9.48)	(6.86)
40.00	20.52	40.00	(30.00)	-	(9.48)	40.00	(6.86)	(9.48)	(6.86)
45.00	15.52	45.00	(25.00)	-	(9.48)	45.00	(6.86)	(9.48)	(6.86)
50.00	10.52	50.00	(20.00)	-	(9.48)	50.00	(6.86)	(9.48)	(6.86)
55.00	5.52	55.00	(15.00)	-	(9.48)	55.00	(6.86)	(9.48)	(6.86)
60.00	0.52	60.00	(10.00)	-	(9.48)	60.00	(6.86)	(9.48)	(6.86)
65.00	(4.48)	65.00	(5.00)	-	(9.48)	65.00	(6.86)	(9.48)	(6.86)
70.00	(9.48)	70.00	-	-	(9.48)	70.00	(6.86)	(9.48)	(6.86)
75.00	(9.48)	75.00	5.00	-	(4.48)	75.00	(1.86)	(4.48)	(1.86)
80.00	(9.48)	80.00	10.00	-	0.52	80.00	3.14	0.52	3.14
85.00	(9.48)	85.00	15.00	-	5.52	85.00	8.14	5.52	8.14
90.00	(9.48)	90.00	20.00	-	10.52	90.00	13.14	10.52	13.14
95.00	(9.48)	95.00	25.00	-	15.52	95.00	18.14	15.52	18.14
100.00	(9.48)	100.00	30.00	-	20.52	100.00	23.14	20.52	23.14
105.00	(9.48)	105.00	35.00	-	25.52	105.00	28.14	25.52	28.14
110.00	(9.48)	110.00	40.00	-	30.52	110.00	33.14	30.52	33.14
115.00	(9.48)	115.00	45.00	-	35.52	115.00	38.14	35.52	38.14
120.00	(9.48)	120.00	50.00	-	40.52	120.00	43.14	40.52	43.14
125.00	(9.48)	125.00	55.00	-	45.52	125.00	48.14	45.52	48.14
130.00	(9.48)	130.00	60.00	-	50.52	130.00	53.14	50.52	53.14
135.00	(9.48)	135.00	65.00	-	55.52	135.00	58.14	55.52	58.14
140.00	(9.48)	140.00	70.00	-	60.52	140.00	63.14	60.52	63.14

Fig. 4.10 (1) Congruence of Protective Put, as well as long Pharma Group share and long put, (2) put-call parity

financial engineer owns the share and is uncertain about the short- to medium-term further share price changes.

The collar is a combination of a covered call and a protective put. Here, the investor is the owner of the stock itself, sells the same number of OTM calls, and buys OTM puts with the same maturity. The profit and loss potential of this strategy is limited. The long put acts as a kind of stop/loss. It protects the financial engineer against larger losses if the share price of the Pharma Group falls sharply. The short call finances the whole thing by receiving the premium. On the other hand, the financial engineer forgoes additional price gains above the strike of the call (72 EUR).

Important Formulas

$$\text{Maximum profit} = \text{strike price short call} - \text{purchase price of the underlying} \\ + \text{premium received} - \text{premium paid}$$

$$\text{Break} - \text{even} = \text{purchase price of the underlying} + \text{premium paid} - \text{premium received}$$

$$\text{Maximum loss} = \text{strike price long put} - \text{purchase price of the underlying} \\ + \text{premium received} - \text{premium paid}$$

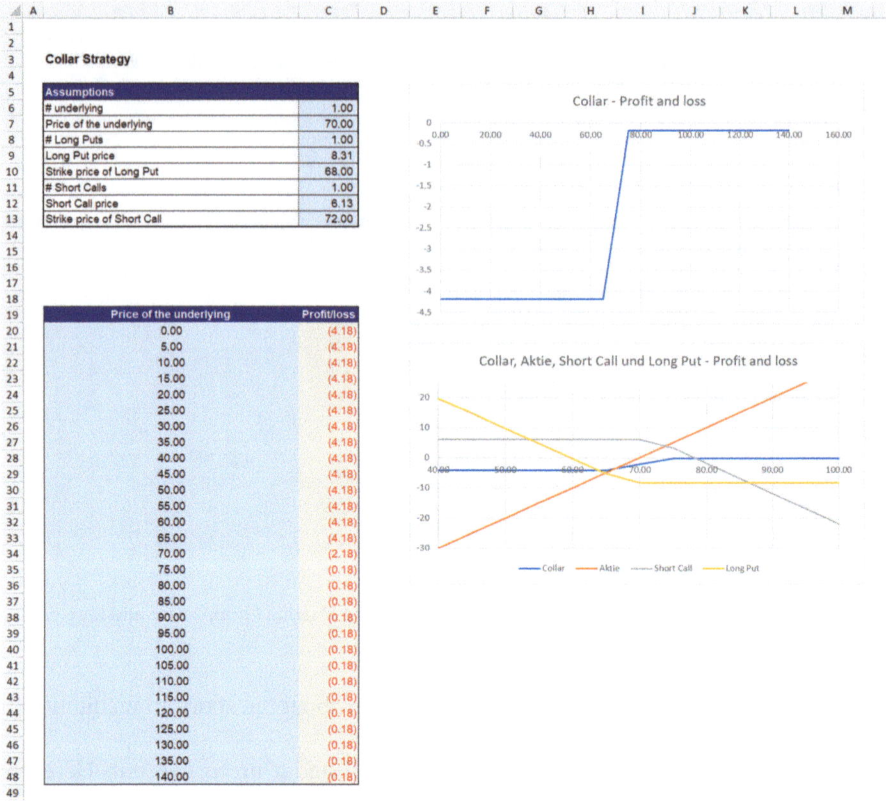

Collar Strategy

Assumptions	
# underlying	1.00
Price of the underlying	70.00
# Long Puts	1.00
Long Put price	8.31
Strike price of Long Put	68.00
# Short Calls	1.00
Short Call price	6.13
Strike price of Short Call	72.00

Price of the underlying	Profit/loss
0.00	(4.18)
5.00	(4.18)
10.00	(4.18)
15.00	(4.18)
20.00	(4.18)
25.00	(4.18)
30.00	(4.18)
35.00	(4.18)
40.00	(4.18)
45.00	(4.18)
50.00	(4.18)
55.00	(4.18)
60.00	(4.18)
65.00	(4.18)
70.00	(2.18)
75.00	(0.18)
80.00	(0.18)
85.00	(0.18)
90.00	(0.18)
95.00	(0.18)
100.00	(0.18)
105.00	(0.18)
110.00	(0.18)
115.00	(0.18)
120.00	(0.18)
125.00	(0.18)
130.00	(0.18)
135.00	(0.18)
140.00	(0.18)

Fig. 4.11 Collar calculation

Display in Excel

You can enter the following formula in cell C20:

=SUM (((B20-$C$7)*$C$6);IF(B20<$C$10;$C$10-B20-$C$9;-$C$9)*$C$8; IF(B20>$C$13;$C$13-B20+$C$12;$C$12)*$C$11)

Following this, you can drag the cell C20 to C48 with the mouse. You will then get the following picture as shown in Fig. 4.11.

Note that the collar only gives an exclusively negative result due to the pricing of the selected options!

The formulas for P20, S20, V20, X20, and Y20 are as follows:

P20: =O20-$P$7
S20 =IF(R20>$S$10;$S$10-$R20+$S$9;$S$9)
V20 =IF(U20<$V$10;$V$10-$U20-$V$9;-$V$9)
X20 =ROUND((C20-Y20);2)
Y20 =P20+S20+V20

Pharma Group share

Assumptions	
Price of the underlying	70.00

Short Call

Assumptions	
Price of the underlying	70.00
Short Call price	6.13
Strike price of Short Call	72.00

Long Put

Assumptions	
Price of the underlying	70.00
Long Put price	8.31
Strike price of Long Put	68.00

Collar
versus
Long share +Short Call+Long Put

Price of the underlying	Profit/loss	Price of the underlying	Profit/loss	Price of the underlying	Profit/loss	Difference	Profit/loss
0.00	(70.00)	0.00	6.13	0.00	59.69	-	(4.18)
5.00	(65.00)	5.00	6.13	5.00	54.69	-	(4.18)
10.00	(60.00)	10.00	6.13	10.00	49.69	-	(4.18)
15.00	(55.00)	15.00	6.13	15.00	44.69	-	(4.18)
20.00	(50.00)	20.00	6.13	20.00	39.69	-	(4.18)
25.00	(45.00)	25.00	6.13	25.00	34.69	-	(4.18)
30.00	(40.00)	30.00	6.13	30.00	29.69	-	(4.18)
35.00	(35.00)	35.00	6.13	35.00	24.69	-	(4.18)
40.00	(30.00)	40.00	6.13	40.00	19.69	-	(4.18)
45.00	(25.00)	45.00	6.13	45.00	14.69	-	(4.18)
50.00	(20.00)	50.00	6.13	50.00	9.69	-	(4.18)
55.00	(15.00)	55.00	6.13	55.00	4.69	-	(4.18)
60.00	(10.00)	60.00	6.13	60.00	(0.31)	-	(4.18)
65.00	(5.00)	65.00	6.13	65.00	(5.31)	-	(4.18)
70.00	-	70.00	6.13	70.00	(8.31)	-	(2.18)
75.00	5.00	75.00	3.13	75.00	(8.31)	-	(0.18)
80.00	10.00	80.00	(1.87)	80.00	(8.31)	-	(0.18)
85.00	15.00	85.00	(6.87)	85.00	(8.31)	-	(0.18)
90.00	20.00	90.00	(11.87)	90.00	(8.31)	-	(0.18)
95.00	25.00	95.00	(16.87)	95.00	(8.31)	-	(0.18)
100.00	30.00	100.00	(21.87)	100.00	(8.31)	-	(0.18)
105.00	35.00	105.00	(26.87)	105.00	(8.31)	-	(0.18)
110.00	40.00	110.00	(31.87)	110.00	(8.31)	-	(0.18)
115.00	45.00	115.00	(36.87)	115.00	(8.31)	-	(0.18)
120.00	50.00	120.00	(41.87)	120.00	(8.31)	-	(0.18)
125.00	55.00	125.00	(46.87)	125.00	(8.31)	-	(0.18)
130.00	60.00	130.00	(51.87)	130.00	(8.31)	-	(0.18)
135.00	65.00	135.00	(56.87)	135.00	(8.31)	-	(0.18)
140.00	70.00	140.00	(61.87)	140.00	(8.31)	-	(0.18)

Fig. 4.12 Congruence of collar and share, short call, and long put

After that, you can drag the respective cell with the mouse to P48, T48, W48, as well as V48. You will then get the following picture as shown in Fig. 4.12.

4.11 Assignment 11: Bullish–Bull Call Spread

Task

(a) How is a bull call spread formed, and what is the loss or profit potential of this strategy?
(b) Calculate the result of a bull call spread strategy when buying an ITM call and selling an OTM call. Possible developments of the Pharma Group share range from 0.00 EUR to 140.00 EUR in steps of five. Plot the result graphically.
(c) Show which elements make up the Bull Call Spread and how the individual values of the Bull Call Spread can be determined for the respective different share prices of the Pharma Group share.

Content

The construction of a bull call spread consists of buying calls and selling the same number of calls with a higher strike price at the same maturity on the same underlying.

By selling the call with a higher strike (here: 72 EUR), the financial engineer deprives himself of the opportunity to profit from a strong increase in the underlying. However, this lowers the cost of his own bullish position. Since the price of the sold call (6.13 EUR) at fair valuation is lower than the price paid for the purchased call (7.67 EUR), the financial engineer enters this position with a loss; i.e., the strategy always requires an initial investment. Therefore, the Bull Call Spread is also called a Bull Call *Debit* Spread. The profit and loss potential of this strategy is limited. In this case, the maximum loss is 1.54 EUR, and the maximum profit is 2.46 EUR. The share price of 69.54 EUR (68+7.67−6.13) separates the profit zone from the loss zone.

Important Formulas

$$\begin{aligned}\text{Maximum profit} &= \text{strike price short call} - \text{strike price long call}\\ &\quad -\text{debit}\left(= \text{premium received} - \text{premium paid}\right)\end{aligned}$$

$$\text{Break} - \text{even} = \text{base price of the long call} + \text{premium paid} - \text{premium received}$$

$$\text{Maximum loss} = \text{debit}$$

Display in Excel
You can enter the following formula in cell C20:

=SUM((IF(B20>$C$8;B20-$C$8-$C$7;-$C$7)*$C$6);IF(B20>$C$11;$C$11-B20+$C$10;$C$10))

Following this, you can drag the cell C20 to C48 with the mouse. You will then get the following picture as shown in Fig. 4.13.

The formulas for P20, S20, U20, and V20 are as follows:

P20:	=IF(O20>$P$10;$P$10-$O20+$P$9;$P$9)
S20	=IF(R20>$S$10;R20-$S$10-$S$9;-$S$9)
U20	=C20-V20
V20	=P20+S20

Subsequently, you can drag the respective cell with the mouse up to row 48. You will then get the following picture as shown in Fig. 4.14.

4.12 Assignment 12: Bullish–Bull Put Spread

Task

(a) How is a Bull Put Spread formed, and why is it a bullish strategy, although, according to the name, this strategy is formed with puts? What is the loss or profit potential of this strategy?

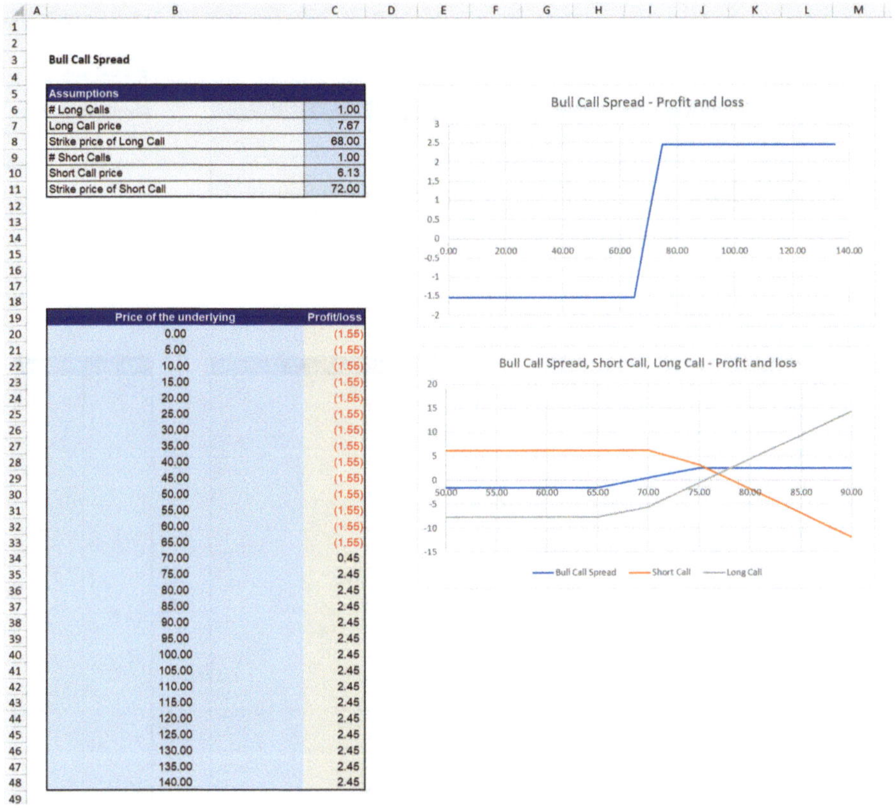

Bull Call Spread

Assumptions	
# Long Calls	1.00
Long Call price	7.67
Strike price of Long Call	68.00
# Short Calls	1.00
Short Call price	6.13
Strike price of Short Call	72.00

Price of the underlying	Profit/loss
0.00	(1.55)
5.00	(1.55)
10.00	(1.55)
15.00	(1.55)
20.00	(1.55)
25.00	(1.55)
30.00	(1.55)
35.00	(1.55)
40.00	(1.55)
45.00	(1.55)
50.00	(1.55)
55.00	(1.55)
60.00	(1.55)
65.00	(1.55)
70.00	0.45
75.00	2.45
80.00	2.45
85.00	2.45
90.00	2.45
95.00	2.45
100.00	2.45
105.00	2.45
110.00	2.45
115.00	2.45
120.00	2.45
125.00	2.45
130.00	2.45
135.00	2.45
140.00	2.45

Fig. 4.13 Calculation of Bull Call Spread

(b) Calculate the result of a bull put spread strategy when buying an OTM put and selling an ITM put. The possible developments of the Pharma Group share range from 0.00 EUR to 140.00 EUR in steps of five. Plot the result graphically.

(c) Show which elements make up the Bull Put Spread and how the individual values of the Bull Put Spread can be determined for the respective different share prices of the Pharma Group share.

Content

The put spread is formed with puts. Although consisting of puts, the Bull Put Spread is nevertheless a bullish strategy. This is due to the fact that the construction of the Bull Put Spread consists of the sale of puts with a higher strike price (here: 72 EUR) and the purchase of the same number of puts (here: 1) with a lower strike price (here: 68 EUR), with the same maturity and on the identical underlying (here: 70 EUR). If the options are fairly valued, the net income is the premiums collected (here: EUR 10.72) less the premiums paid (here: EUR 8.31). The bull put spread, therefore, counts as a *credit* spread. Similar to the Bull Call Spread, the Bull Put Spread also has a limited profit and loss profile.

Short Call

Assumptions	
Price of the underlying	70.00
Short Call price	6.13
Strike price of Short Call	72.00

Long Call

Assumptions	
Price of the underlying	70.00
Long Call price	7.67
Strike price of Long Call	68.00

Bull Call spread
versus
Short Call+Long Call

Price of the underlying	Profit/loss	Price of the underlying	Profit/loss	Difference	Profit/loss
0.00	6.13	0.00	(7.67)	-	(1.55)
5.00	6.13	5.00	(7.67)	-	(1.55)
10.00	6.13	10.00	(7.67)	-	(1.55)
15.00	6.13	15.00	(7.67)	-	(1.55)
20.00	6.13	20.00	(7.67)	-	(1.55)
25.00	6.13	25.00	(7.67)	-	(1.55)
30.00	6.13	30.00	(7.67)	-	(1.55)
35.00	6.13	35.00	(7.67)	-	(1.55)
40.00	6.13	40.00	(7.67)	-	(1.55)
45.00	6.13	45.00	(7.67)	-	(1.55)
50.00	6.13	50.00	(7.67)	-	(1.55)
55.00	6.13	55.00	(7.67)	-	(1.55)
60.00	6.13	60.00	(7.67)	-	(1.55)
65.00	6.13	65.00	(7.67)	-	(1.55)
70.00	6.13	70.00	(5.67)	-	0.45
75.00	3.13	75.00	(0.67)	-	2.45
80.00	(1.87)	80.00	4.33	-	2.45
85.00	(6.87)	85.00	9.33	-	2.45
90.00	(11.87)	90.00	14.33	-	2.45
95.00	(16.87)	95.00	19.33	-	2.45
100.00	(21.87)	100.00	24.33	-	2.45
105.00	(26.87)	105.00	29.33	-	2.45
110.00	(31.87)	110.00	34.33	-	2.45
115.00	(36.87)	115.00	39.33	-	2.45
120.00	(41.87)	120.00	44.33	-	2.45
125.00	(46.87)	125.00	49.33	-	2.45
130.00	(51.87)	130.00	54.33	-	2.45
135.00	(56.87)	135.00	59.33	-	2.45
140.00	(61.87)	140.00	64.33	-	2.45

Fig. 4.14 Congruence of Bull Call Spread with short call and long call

Important Formulas

$$\text{Maximum profit} = \text{credit}\left(\text{premium received} - \text{premium paid}\right)$$

$$\text{Break} - \text{even} = \text{strike price of the short put credit}$$

$$\text{Maximum loss} = \text{strike price long put} - \text{strike price short put} + \text{credit}$$

Display in Excel
You can enter the following formula in cell C20:

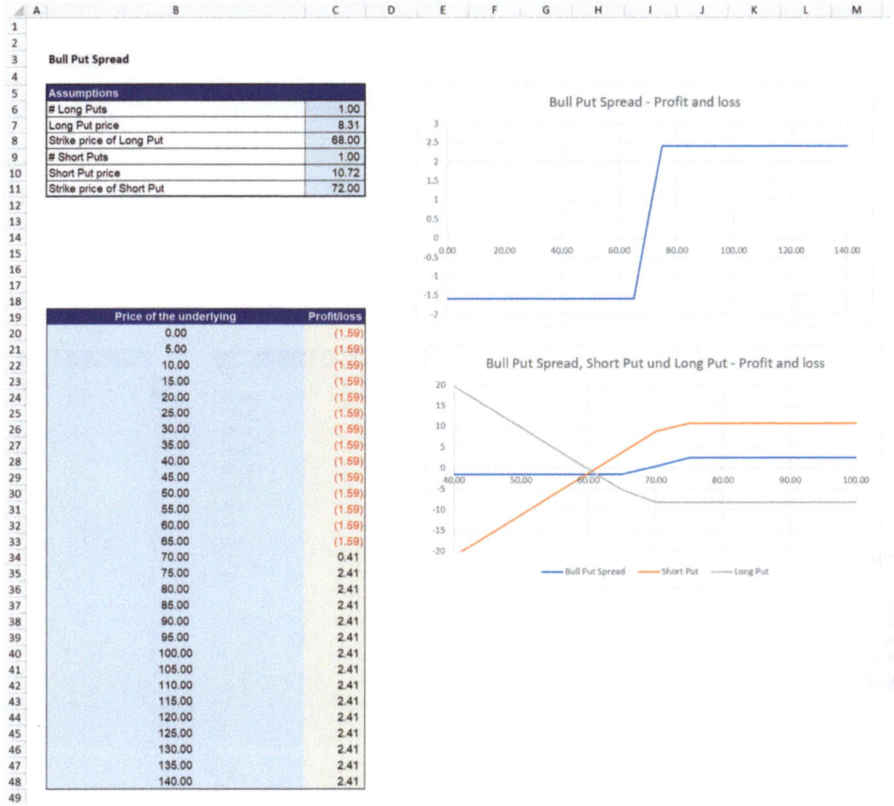

Bull Put Spread

Assumptions	
# Long Puts	1.00
Long Put price	8.31
Strike price of Long Put	68.00
# Short Puts	1.00
Short Put price	10.72
Strike price of Short Put	72.00

Price of the underlying	Profit/loss
0.00	(1.59)
5.00	(1.59)
10.00	(1.59)
15.00	(1.59)
20.00	(1.59)
25.00	(1.59)
30.00	(1.59)
35.00	(1.59)
40.00	(1.59)
45.00	(1.59)
50.00	(1.59)
55.00	(1.59)
60.00	(1.59)
65.00	(1.59)
70.00	0.41
75.00	2.41
80.00	2.41
85.00	2.41
90.00	2.41
95.00	2.41
100.00	2.41
105.00	2.41
110.00	2.41
115.00	2.41
120.00	2.41
125.00	2.41
130.00	2.41
135.00	2.41
140.00	2.41

Fig. 4.15 Calculation of Bull Put Spread

=SUM(IF(B20<$C$8;$C$8-B20-$C$7;-$C$7)*$C$6;IF(B20<$C$11;B20-$C$11+$C$10;$C$10)*$C$9)

Following this, you can drag the cell C20 to C48 with the mouse. You will then get the following picture as shown in Fig. 4.15.

The formulas for P20, S20, U20, and V20 are as follows:

P20	=IF($O20<$P$11;$P$10-$P$11+$O20;$P$10)
S20	=IF(R20<$S$8;$S$8-$R20-$S$7;-$S$7)
U20	=ROUND((C20-V20);2)
V20	=P20+S20

Subsequently, you can drag the respective cell with the mouse up to row 48. You will then get the following picture as shown in Fig. 4.16.

Short Put

Assumptions	
Short Put price	10.72
Strike price of Short Put	72.00

Long Put

Assumptions	
Long Put price	8.31
Strikeprice of Long Put	68.00

Bull Put Spread
versus
Short Put+Long Put

Price of the underlying	Profit/loss	Price of the underlying	Profit/loss	Difference	Profit/loss
0.00	(61.28)	0.00	59.69	-	(1.59)
5.00	(56.28)	5.00	54.69	-	(1.59)
10.00	(51.28)	10.00	49.69	-	(1.59)
15.00	(46.28)	15.00	44.69	-	(1.59)
20.00	(41.28)	20.00	39.69	-	(1.59)
25.00	(36.28)	25.00	34.69	-	(1.59)
30.00	(31.28)	30.00	29.69	-	(1.59)
35.00	(26.28)	35.00	24.69	-	(1.59)
40.00	(21.28)	40.00	19.69	-	(1.59)
45.00	(16.28)	45.00	14.69	-	(1.59)
50.00	(11.28)	50.00	9.69	-	(1.59)
55.00	(6.28)	55.00	4.69	-	(1.59)
60.00	(1.28)	60.00	(0.31)	-	(1.59)
65.00	3.72	65.00	(5.31)	-	(1.59)
70.00	8.72	70.00	(8.31)	-	0.41
75.00	10.72	75.00	(8.31)	-	2.41
80.00	10.72	80.00	(8.31)	-	2.41
85.00	10.72	85.00	(8.31)	-	2.41
90.00	10.72	90.00	(8.31)	-	2.41
95.00	10.72	95.00	(8.31)	-	2.41
100.00	10.72	100.00	(8.31)	-	2.41
105.00	10.72	105.00	(8.31)	-	2.41
110.00	10.72	110.00	(8.31)	-	2.41
115.00	10.72	115.00	(8.31)	-	2.41
120.00	10.72	120.00	(8.31)	-	2.41
125.00	10.72	125.00	(8.31)	-	2.41
130.00	10.72	130.00	(8.31)	-	2.41
135.00	10.72	135.00	(8.31)	-	2.41
140.00	10.72	140.00	(8.31)	-	2.41

Fig. 4.16 Congruence of Bull Put Spread with short put and long put

4.13 Assignment 13: Bullish–Call Backspread

Task

(a) How is a call backspread formed, and what is the loss or profit potential of this strategy?

(b) Calculate the result of a call backspread strategy when buying two OTM calls and selling one ITM call. The possible developments of the Pharma Group share range from 0.00 EUR to 140.00 EUR in steps of five. Plot the result graphically.

(c) Show which elements make up the call backspread and how the individual values of the call backspread can be determined for the respective different share prices of the Pharma Group share.

Content

The call backspread strategy is formed by the investor buying more calls (here: 2) with a higher strike price (here: 72 EUR) than he sells with a lower strike price (here: 68 EUR). The underlying asset (here: 70 EUR) and the terms (here: 1 year) of the options are identical. The financial engineer of such a strategy assumes a substantial price gain of the underlying. This strategy has an unlimited profit and a limited loss potential. The call backspread shown here is thus formed by the purchase of two calls and the sale of one call. It is therefore similar to the option strategy "Long Call" (see assignment 2). However, the two strategies differ in that the maximum profit or loss is generally lower for the call backspread than for the long call.

1. At a share price of 0, the "maximum loss" here is EUR 4.59 (instead of EUR 7.67). The maximum gain at a share price of 140 EUR is 59.41 EUR (instead of 64.33 EUR).
2. Restricting point 1, there is an even larger loss on the call backspread at the strike of the long call (here: 72 EUR). The maximum loss here is 8.59 EUR (instead of a constant −7.67 EUR).

Important Formulas

$$\text{Profit} = \text{price of the underlying asset} - \text{strike price long call} - \text{maximum loss}$$

$$\text{Break} - \text{even} = \text{strike price long call} - \text{maximum loss}$$

If the premium paid and the premium received balance each other out, this results in a second, lower break-even at the strike price of the short call.

$$\text{Maximum loss} = \text{strike price short call} - \text{strike price long call} + \text{premium received} \\ -\text{premium paid}$$

Display in Excel

You can enter the following formula in cell C20:

=SUM((IF(B20>$C$8;B20-$C$8-$C$7;-$C$7)*$C$6);IF(B20>$C$11;$C$11-B20+$C$10;$C$10))

Following this, you can drag the cell C20 to C48 with the mouse. You will then get the following picture as shown in Fig. 4.17.

The formulas for P20, S20, U20, and V20 are as follows:

P20: =IF(O20>$P$8;O20-$P$8-$P$7;-$P$7)*$P$6
S20 =IF(R20>$S$11;$S$11-$R20+$S$10;$S$10)*$S$9
U20 =C20-V20
V20 =P20+S20

Subsequently, you can drag the respective cell with the mouse up to row 48. You will then get the following picture as shown in Fig. 4.18.

Call Backspread

Assumptions	
# Long Calls	2.00
Long Call price	6.13
Strike price of Long Call	72.00
# Short Calls	1.00
Short Call price	7.67
Strike price of Short Call	68.00

Price of the underlying	Profit/loss
0.00	(4.58)
5.00	(4.58)
10.00	(4.58)
15.00	(4.58)
20.00	(4.58)
25.00	(4.58)
30.00	(4.58)
35.00	(4.58)
40.00	(4.58)
45.00	(4.58)
50.00	(4.58)
55.00	(4.58)
60.00	(4.58)
65.00	(4.58)
70.00	(6.58)
75.00	(5.58)
80.00	(0.58)
85.00	4.42
90.00	9.42
95.00	14.42
100.00	19.42
105.00	24.42
110.00	29.42
115.00	34.42
120.00	39.42
125.00	44.42
130.00	49.42
135.00	54.42
140.00	59.42

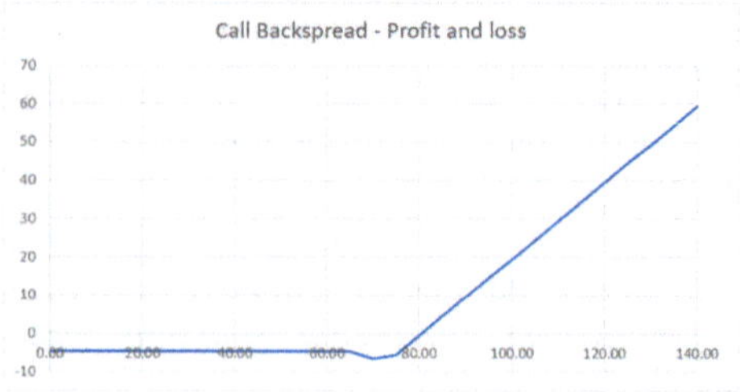

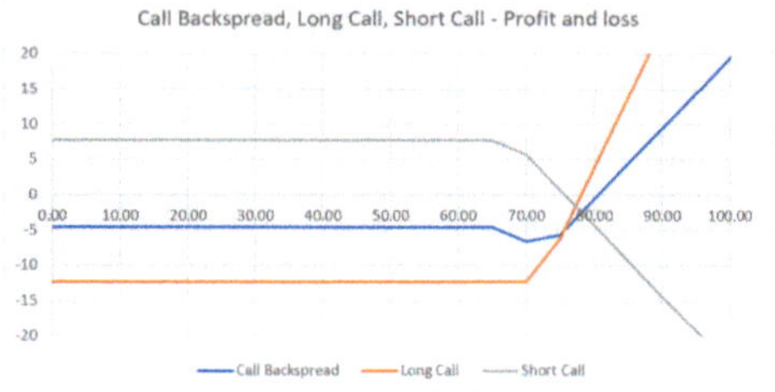

Fig. 4.17 Calculation of call backspread

Long Call

Assumptions	
# Long Calls	2.00
Long Call price	6.13
Strike price of Long Call	72.00

Price of the underlying	Profit/loss
0.00	(12.25)
5.00	(12.25)
10.00	(12.25)
15.00	(12.25)
20.00	(12.25)
25.00	(12.25)
30.00	(12.25)
35.00	(12.25)
40.00	(12.25)
45.00	(12.25)
50.00	(12.25)
55.00	(12.25)
60.00	(12.25)
65.00	(12.25)
70.00	(12.25)
75.00	(6.25)
80.00	3.75
85.00	13.75
90.00	23.75
95.00	33.75
100.00	43.75
105.00	53.75
110.00	63.75
115.00	73.75
120.00	83.75
125.00	93.75
130.00	103.75
135.00	113.75
140.00	123.75

Short Call

Assumptions	
# Short Calls	1.00
Short Call price	7.67
Strike price of Short Call	68.00

Price of the underlying	Profit/loss
0.00	7.67
5.00	7.67
10.00	7.67
15.00	7.67
20.00	7.67
25.00	7.67
30.00	7.67
35.00	7.67
40.00	7.67
45.00	7.67
50.00	7.67
55.00	7.67
60.00	7.67
65.00	7.67
70.00	5.67
75.00	0.67
80.00	(4.33)
85.00	(9.33)
90.00	(14.33)
95.00	(19.33)
100.00	(24.33)
105.00	(29.33)
110.00	(34.33)
115.00	(39.33)
120.00	(44.33)
125.00	(49.33)
130.00	(54.33)
135.00	(59.33)
140.00	(64.33)

Call Backspread
versus
Long Call + Short Call

Difference	Profit/loss
-	(4.58)
-	(4.58)
-	(4.58)
-	(4.58)
-	(4.58)
-	(4.58)
-	(4.58)
-	(4.58)
-	(4.58)
-	(4.58)
-	(4.58)
-	(4.58)
-	(4.58)
-	(4.58)
-	(6.58)
-	(5.58)
-	(0.58)
-	4.42
-	9.42
-	14.42
-	19.42
-	24.42
-	29.42
-	34.42
-	39.42
-	44.42
-	49.42
-	54.42
-	59.42

Fig. 4.18 Congruence of call backspread and long call and short call

Further Reading

Ernst, Dietmar; Häcker, Joachim (2016): Financial Modeling, pp. 801–810.
Hull, John (2015): Options, futures, and other derivatives. S. 325–348.
See Excel file Case Study Derivatives Course 2, Excel worksheet "Input," "Derivation of Option Prices," and 3–6.
Toolbox, Tasks 228–283.
See Basic Option Strategies Video.
Ernst, Dietmar; Häcker, Joachim (2016): Financial Modeling, pp. 802–804.
See Excel file Case Study Derivatives Part 2, Excel worksheet 3.
Toolbox, Tasks 231–234.
See Basic Option Strategies video.
Ernst, Dietmar; Häcker, Joachim (2016): Financial Modeling, pp. 804–807.
See Excel file Case Study Derivatives Part 2, Excel worksheet 4.
Toolbox, Tasks 239–241.

See Basic Option Strategies Video.
Ernst, Dietmar; Häcker, Joachim (2016): Financial Modeling, pp. 807–808.
See Excel file Case Study Derivatives Part 2, Excel worksheet 5.
Toolbox, Tasks 235–238.
See Basic Option Strategies Video.
Ernst, Dietmar; Häcker, Joachim (2016): Financial Modeling, pp. 809–810.
See Excel file Case Study Derivatives Part 2, Excel worksheet 6.
Toolbox, Tasks 242–244.
See Basic Option Strategies Video
See Excel file Case Study Derivatives Part 2, Excel worksheet 8–40.
Toolbox, Tasks 228–283.
www.theoptionsguide.com/option-trading-strategies.aspx
Hull, John (2015): Options, futures, and other derivatives. S. 325–348.
Rieger, M.O. (2016): Options, Derivatives and Structured Products, pp. 113–160.
See video "Advanced Option Strategies – an overview."
Ernst, Dietmar; Häcker, Joachim (2016): Financial Modeling, p. 806.
See Excel file Case Study Derivatives Part 2, Excel worksheet 8–9.
See video "Covered Call OTM."
See Excel file Case Study Derivatives Part 2, Excel worksheet 10–11.
See video "Protective Put."
See Excel file Case Study Derivatives Part 2, Excel worksheet 12–13.
See Excel file Case Study Derivatives Part 2, Excel worksheet 14–15.
See Excel file Case Study Derivatives Part 2, Excel worksheet 16–17.
See Excel file Case Study Derivatives Part 2, Excel worksheet 18–19.
See Call Backspread video.

Chapter 5
Bearish Advanced Options Strategies

5.1 Assignment 14: Bearish–Covered Put

Task

(a) How is a covered put formed, and what is the loss or profit potential of this strategy?
(b) Calculate the result of a covered put strategy when selling an ATM put and an underlying. Possible developments of the Pharma Group share range from 0.00 EUR to 140.00 EUR in steps of five. Plot the result graphically.

Content
The covered put is a form of a synthetic short call. Here, the investor enters into a short position in the underlying itself while selling a put. In the event of a rise in the underlying, the investor has unlimited loss potential. The profit potential is limited and restricted to the premium of the sold put.

Important Formulas

$$\text{Maximum profit if} = \text{price of the underlying} <= \text{strike price of the put}$$

$$\text{Break} - \text{even} = \text{selling price of the underlying} + \text{premium received}$$

$$\text{Loss} = \text{selling price of the underlying} + \text{premium received} - \text{price of the underlying}$$

Display in Excel
You can enter the following formula in cell C20:

=SUM((C$7-B20)*$C$6;IF(B20<$C$10;B20-$C$10+$C$9;$C$9)*$C$8)

Supplementary Information The online version contains supplementary material available at https://doi.org/10.1007/978-3-031-85822-2_5.

D. Ernst, J. Häcker, *Derivatives*, Springer Texts in Business and Economics,
https://doi.org/10.1007/978-3-031-85822-2_5

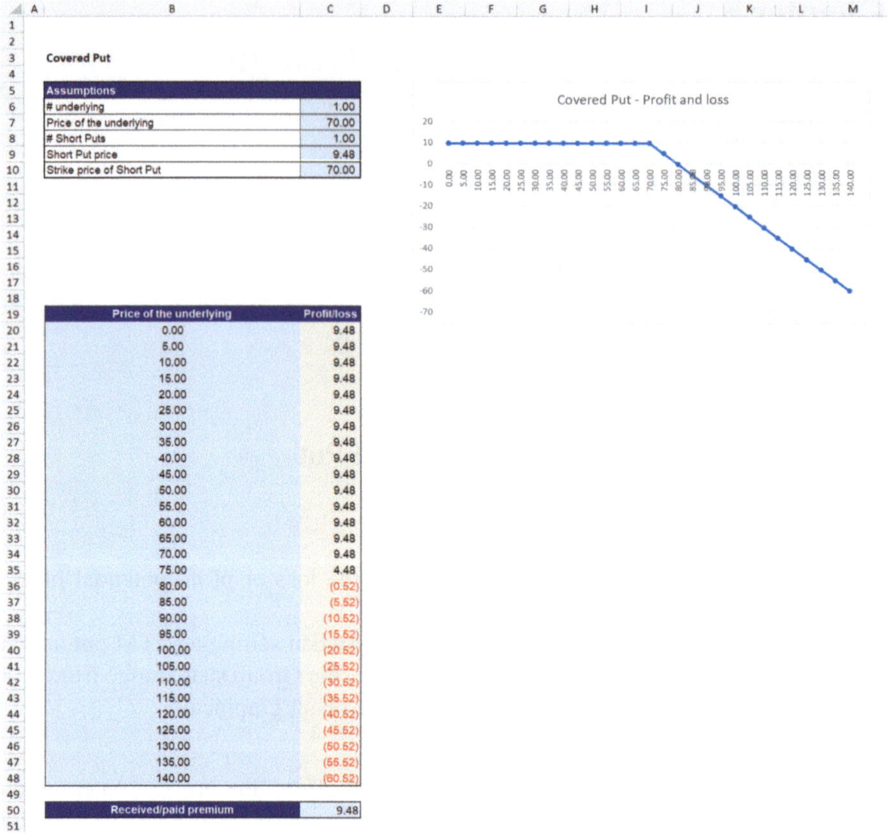

Assumptions	
# underlying	1.00
Price of the underlying	70.00
# Short Puts	1.00
Short Put price	9.48
Strike price of Short Put	70.00

Price of the underlying	Profit/loss
0.00	9.48
5.00	9.48
10.00	9.48
15.00	9.48
20.00	9.48
25.00	9.48
30.00	9.48
35.00	9.48
40.00	9.48
45.00	9.48
50.00	9.48
55.00	9.48
60.00	9.48
65.00	9.48
70.00	9.48
75.00	4.48
80.00	(0.52)
85.00	(5.52)
90.00	(10.52)
95.00	(15.52)
100.00	(20.52)
105.00	(25.52)
110.00	(30.52)
115.00	(35.52)
120.00	(40.52)
125.00	(45.52)
130.00	(50.52)
135.00	(55.52)
140.00	(60.52)

Received/paid premium	9.48

Fig. 5.1 Calculation of covered put

Following this, you can drag the cell C20 to C48 with the mouse. You will then get the following picture as shown in Fig. 5.1.

5.2 Assignment 15: Bearish–Put Backspread

Task

(a) How is a put backspread formed, and what is the loss or profit potential of this strategy?
(b) Calculate the result of a put backspread strategy when selling an ITM put and buying two OTM puts. Possible developments of the Pharma Group share range from 0.00 EUR to 140.00 EUR in steps of five. Plot the result graphically.

Content

The put backspread is the counterpart of the call backspread. This strategy is formed by selling a put option with strike price B and buying more put options with strike price A. The strike price B is usually at the current share price, and the strike price A is usually below the current share price.

Important Formulas

$$\text{Profit} = \text{strike price long put} - \text{price of the underlying asset} - \text{maximum loss}$$

$$\text{Break even} = \text{strike price of the long put} - \text{maximum loss}$$

Provided that the premium paid and the premium received balance each other out, there is a second, higher break-even at the strike price of the short put.

$$\text{Maximum loss} = \text{strike price long put} - \text{strike price short put} + \text{premium received} - \text{premium paid}$$

Display in Excel

You can enter the following formula in cell C20:

```
=SUM((IF(B20<$C$8;$C$8-B20-$C$7;-$C$7)*$C$6);IF(B20<$C$11;B20-
$C$11+$C$10;$C$10)*$C$9)
```

Following this, you can drag the cell C20 to C48 with the mouse. You will then get the following picture as shown in Fig. 5.2.

5.3 Assignment 16: Bearish–Bear Put Spread

Task

(a) How is a Bear Put Spread formed, and what is the loss or profit potential of this strategy?
(b) Calculate the result of a bear put spread strategy when selling an OTM put and buying an ITM put. Possible developments of the Pharma Group share range from 0.00 EUR to 140.00 EUR in steps of five. Plot the result graphically.

Content

Similar to the Bull Put Spread, in the Bear Put Spread, put options are bought in a quantitatively balanced and sold at the same time. It should be noted that the strike price of the purchased options is higher than that of the sold options. Therefore, the Bear Put Spread also belongs to the debit spread strategies. Both the profit and the loss potential are limited with this strategy.

Important Formulas

$$\text{Maximum profit} = \text{strike price long put} - \text{strike price short put} - \text{debit} \\ (= \text{premium received} - \text{premium paid})$$

$$\text{Break even} = \text{strike price of the long put} + \text{premium paid} - \text{premium received}$$

$$\text{Maximum loss} = \text{debit}$$

Display in Excel

You can enter the following formula in cell C20:

Put Backspread

Assumptions	
# Long Put	2.00
Long Put price	8.31
Strike price of Long Put	68.00
# Short Puts	1.00
Short Put price	10.72
Strike price of Short Put	72.00

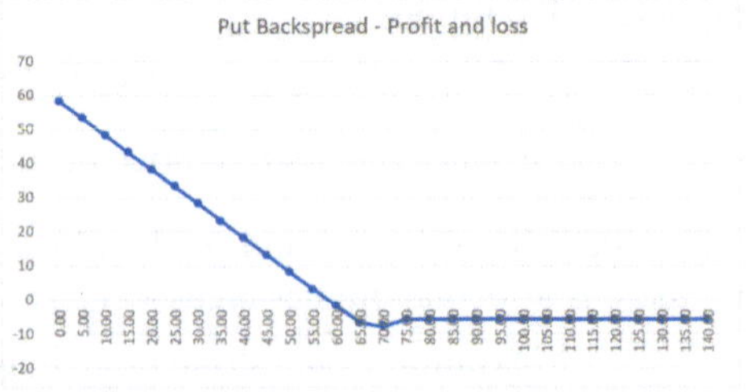

Price of the underlying	Profit/loss
0.00	58.10
5.00	53.10
10.00	48.10
15.00	43.10
20.00	38.10
25.00	33.10
30.00	28.10
35.00	23.10
40.00	18.10
45.00	13.10
50.00	8.10
55.00	3.10
60.00	(1.90)
65.00	(6.90)
70.00	(7.90)
75.00	(5.90)
80.00	(5.90)
85.00	(5.90)
90.00	(5.90)
95.00	(5.90)
100.00	(5.90)
105.00	(5.90)
110.00	(5.90)
115.00	(5.90)
120.00	(5.90)
125.00	(5.90)
130.00	(5.90)
135.00	(5.90)
140.00	(5.90)

Received/paid premium	(5.90)

Fig. 5.2 Calculation of put backspread

=SUM((IF(B20<$C$8;$C$8-B20-$C$7;-$C$7)*$C$6);IF(B20<$C$11;B20-$C$11+$C$10;$C$10)*$C$9)

Following this, you can drag the cell C20 to C48 with the mouse. You will then get the following picture as shown in Fig. 5.3.

5.4 Assignment 17: Bearish–Bear Call Spread

Task

(a) How is a Bear Call Spread formed, and what is the loss or profit potential of this strategy?
(b) Calculate the result of a Bear Call Spread strategy when selling an ITM Call and buying an OTM Call. Possible developments of the Pharma Group share range from 0.00 EUR to 140.00 EUR in steps of five. Plot the result graphically.

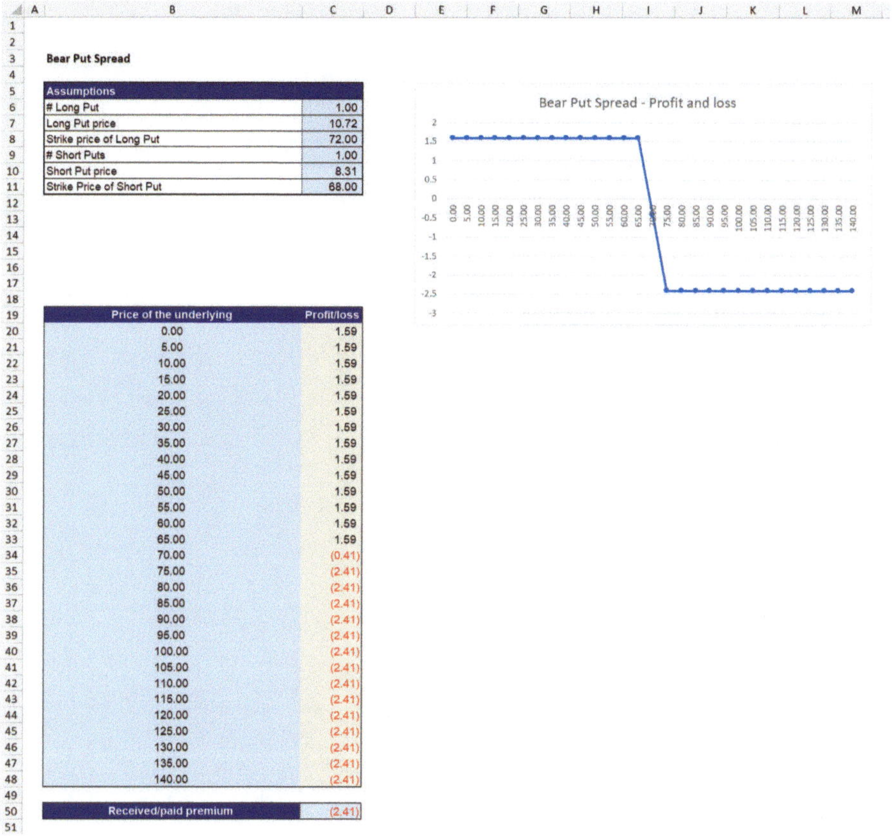

Bear Put Spread

Assumptions	
# Long Put	1.00
Long Put price	10.72
Strike price of Long Put	72.00
# Short Puts	1.00
Short Put price	8.31
Strike Price of Short Put	68.00

Price of the underlying	Profit/loss
0.00	1.59
5.00	1.59
10.00	1.59
15.00	1.59
20.00	1.59
25.00	1.59
30.00	1.59
35.00	1.59
40.00	1.59
45.00	1.59
50.00	1.59
55.00	1.59
60.00	1.59
65.00	1.59
70.00	(0.41)
75.00	(2.41)
80.00	(2.41)
85.00	(2.41)
90.00	(2.41)
95.00	(2.41)
100.00	(2.41)
105.00	(2.41)
110.00	(2.41)
115.00	(2.41)
120.00	(2.41)
125.00	(2.41)
130.00	(2.41)
135.00	(2.41)
140.00	(2.41)

Received/paid premium	(2.41)

Fig. 5.3 Calculation of Bear Put Spread

Content

The Bear Call Spread strategy is the counterpart of the Bull Put Spread. By selling calls with a lower strike price and buying calls with a higher strike price, a positive result is achieved after offsetting the premiums. Thus, the Bear Call Spread is also a credit spread strategy. This strategy has a limited risk of loss and profit.

Important Formulas

$$\text{Maximum profit} = \text{credit}\left(\text{premium received minus premium paid}\right)$$

$$\text{Break even} = \text{strike price of the short call} + \text{premium received} - \text{premium paid}$$

$$\begin{aligned}\text{Maximum loss} = {} & \text{strike price short call} - \text{strike price long call} \\ & + \text{premium received} - \text{premium paid}\end{aligned}$$

Display in Excel

You can enter the following formula in cell C20:

Bear Call Spread

Assumptions	
# Long Calls	1.00
Long Call price	6.13
Strike price of Long Call	72.00
# Short Calls	1.00
Short Call price	7.67
Strike price of Short Call	68.00

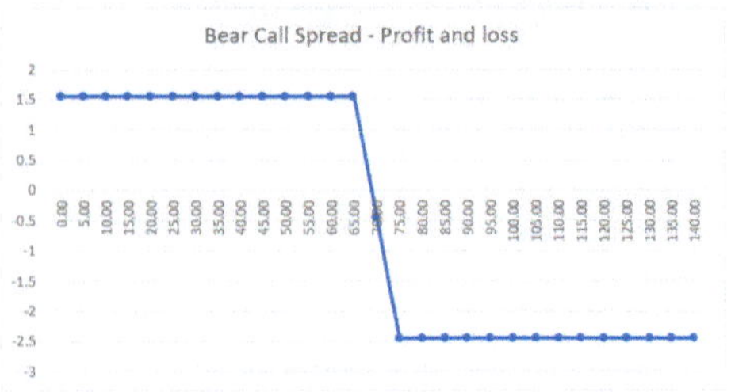

Price of the underlying	Profit/loss
0.00	1.55
5.00	1.55
10.00	1.55
15.00	1.55
20.00	1.55
25.00	1.55
30.00	1.55
35.00	1.55
40.00	1.55
45.00	1.55
50.00	1.55
55.00	1.55
60.00	1.55
65.00	1.55
70.00	(0.45)
75.00	(2.45)
80.00	(2.45)
85.00	(2.45)
90.00	(2.45)
95.00	(2.45)
100.00	(2.45)
105.00	(2.45)
110.00	(2.45)
115.00	(2.45)
120.00	(2.45)
125.00	(2.45)
130.00	(2.45)
135.00	(2.45)
140.00	(2.45)

Received/paid premium	1.55

Fig. 5.4 Calculation of Bear Call Spread

=SUM((IF(B20>$C$8;B20-$C$8-$C$7;-$C$7)*$C$6);IF(B20>$C$11;$C$11-B20+$C$10;$C$10))

Following this, you can drag the cell C20 to C48 with the mouse. You will then get the following picture as shown in Fig. 5.4.

5.5 Assignment 18: Bearish–Protective Call

Task

(a) How is a protective call formed, and what is the loss or profit potential of this strategy?

(b) Calculate the result of a protective call strategy when selling an underlying and buying an ATM call. Possible developments of the Pharma Group share range from 0.00 EUR to 140.00 EUR in steps of five. Plot the result graphically.

Content

A protective call is a form of synthetic long put. This strategy is a combination of selling the underlying and buying call options. This strategy is usually chosen when the investor expects uncertainty in the near future but is still bearish on the underlying. Similar to the Protective Put already explained on the bullish side, the Protective Call is also a typical strategy to hedge profits, in this case from the short position. The profit potential of this strategy is unlimited, and the loss potential is limited.

Important Formulas

$$\text{Profit} = \text{selling price of the underlying} - \text{price of the underlying} - \text{premium paid}$$

$$\text{Break} - \text{even} = \text{selling price of the underlying} + \text{premium paid}$$

$$\text{Maximum loss} = \text{selling price of the underlying} - \text{strike price long call} \\ - \text{paid premium}$$

Display in Excel

You can enter the following formula in cell C20:

=SUM((IF(B20>$C$10;B20-$C$10-$C$9;-$C$9)*$C$8);($C$7-B20)*$C$6)

Following this, you can drag the cell C20 to C48 with the mouse. You will then get the following picture as shown in Fig. 5.5.

5.6 Assignment 19: Neutral Bearish–Condor

Task

(a) How is a Condor Options strategy formed, and what is the loss or profit potential of this strategy?

(b) Calculate the result of a Condor Options strategy when buying an OTM Call (higher strike price) and an ITM Call (lower strike price) while selling an ITM Call and an OTM Call. Possible developments of the Pharma Group share range from 0.00 EUR to 140.00 EUR in steps of five. Plot the result graphically. Please note the following additional information:

Type	Strike	Price (according to Black-Scholes-Merton)
Call (lower strike price)	66.00	8.56
Call (higher strike price)	74.00	5.46

Content

In the Condor Options strategy, the investor assumes that the volatility of the underlying asset will be low. This strategy belongs to the so-called wingspread strategies, as its graphic appearance reminds of a flying creature. The terms and the underlying of the options are identical. So, as with debit spreads, the investor has to spend a

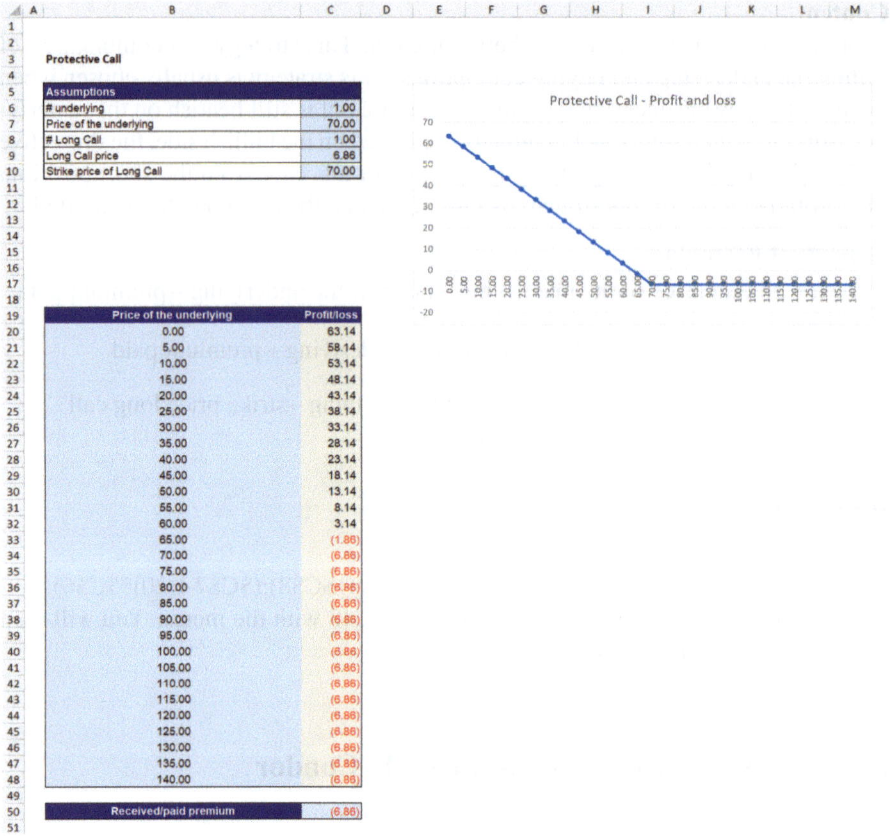

Protective Call

Assumptions	
# underlying	1.00
Price of the underlying	70.00
# Long Call	1.00
Long Call price	6.86
Strike price of Long Call	70.00

Price of the underlying	Profit/loss
0.00	63.14
5.00	58.14
10.00	53.14
15.00	48.14
20.00	43.14
25.00	38.14
30.00	33.14
35.00	28.14
40.00	23.14
45.00	18.14
50.00	13.14
55.00	8.14
60.00	3.14
65.00	(1.86)
70.00	(6.86)
75.00	(6.86)
80.00	(6.86)
85.00	(6.86)
90.00	(6.86)
95.00	(6.86)
100.00	(6.86)
105.00	(6.86)
110.00	(6.86)
115.00	(6.86)
120.00	(6.86)
125.00	(6.86)
130.00	(6.86)
135.00	(6.86)
140.00	(6.86)

Received/paid premium	(6.86)

Fig. 5.5 Protective call calculation

higher premium at the beginning than he receives. The profit and loss potential of this strategy are limited.

Important Formulas

$$\text{Maximum profit} = \text{strike price of the lower short call} - \text{strike price of the lower long call} - \text{premium paid} + \text{premium received}$$

Due to its construction, the Condor Options strategy usually has two break-even points:

$$\text{Higher break-even} = \text{strike price higher long call} - \text{premium paid} + \text{premium received}$$

$$\text{Lower break-even} = \text{strike price lower long call} + \text{premium paid} - \text{premium received}$$

$$\text{Maximum loss} = \text{premium received} - \text{premium paid}$$

Display in Excel
You can enter the following formula in cell C20:

=SUM(IF(B20>$C$8;B20-$C$8-$C$7;-$C$7)*$C$6;IF(B20>$C$11;B20-$C$11-$C$10;-$C$10)*$C$9;IF(B20>$C$14;$C$14-B20+$C$13;$C$13)*$C$12;IF(B20>$C$17;$C$17-B20+$C$16;$C$16)*$C$15)

Following this, you can drag the cell C20 to C48 with the mouse. You will then get the following picture as shown in Fig. 5.6.

5.7 Assignment 20: Neutral Bearish–Long Call Butterfly

Task

(a) How is a Long Call Butterfly strategy formed, and what is the loss or profit potential of this strategy?
(b) Calculate the result of a long call butterfly strategy when selling two ATM calls and buying one ITM and one OTM call. Possible developments of the Pharma Group share range from 0.00 EUR to 140.00 EUR in steps of five. Plot the result graphically.

Content
The Butterfly Spread is the combination of a Bull and a Bear Spread. Therefore, this neutral strategy can be constructed both as a call butterfly and a put butterfly. The long call butterfly variant can be formed by buying one ITM call and one OTM call while selling two ATM calls. Thus, three different strike prices are involved, and the maturities are identical. The profit and loss potential of this strategy are limited.

Important Formulas

$$\text{Maximum profit} = \text{strike price of short calls} - \text{strike price of lower long call} + \text{premium received} - \text{premium paid}$$

$$\text{Higher break-even} = \text{strike price of the higher long call} + \text{premium received} - \text{premium paid}$$

$$\text{Lower break-even} = \text{strike price of lower long call} - \text{premium received} + \text{premium paid}$$

$$\text{Maximum loss} = \text{premium received} - \text{premium paid}$$

Display in Excel
You can enter the following formula in cell C20:

=SUM(IF(B20>$C$8;B20-$C$8-$C$7;-$C$7)*$C$6;IF(B20>$C$11;B20-$C$11-$C$10;-$C$10)*$C$9;IF(B20>$C$14;$C$14-B20+$C$13;$C$13)*$C$12)

Following this, you can drag the cell C20 to C48 with the mouse. You will then get the following picture as shown in Fig. 5.7.

Condor Options

Assumptions	
# Long Calls (OTM higher strike price)	1.00
Long Call price (OTM higher strike price)	5.46
Strike price of Long Call (OTM higher strike price)	74.00
# Long Calls (ITM lower strike price)	1.00
Long Call price (ITM lower strike price)	8.56
Strike price of Long Call (ITM lower strike price)	66.00
# Short Calls (OTM)	1.00
Short Call price (OTM)	6.13
Strike price of Short Call (OTM)	72.00
# Short Calls (ITM)	1.00
Short Call price (ITM)	7.67
Strike price of Short Call (ITM)	68.00

Price of the underlying	Profit/loss
0.00	(0.22)
5.00	(0.22)
10.00	(0.22)
15.00	(0.22)
20.00	(0.22)
25.00	(0.22)
30.00	(0.22)
35.00	(0.22)
40.00	(0.22)
45.00	(0.22)
50.00	(0.22)
55.00	(0.22)
60.00	(0.22)
65.00	(0.22)
70.00	1.78
75.00	(0.22)
80.00	(0.22)
85.00	(0.22)
90.00	(0.22)
95.00	(0.22)
100.00	(0.22)
105.00	(0.22)
110.00	(0.22)
115.00	(0.22)
120.00	(0.22)
125.00	(0.22)
130.00	(0.22)
135.00	(0.22)
140.00	(0.22)

Received/paid premium	(0.22)

Fig. 5.6 Calculation of Condor Options

5.8 Assignment 21: Neutral Bearish–Long Put Butterfly

Task

(a) How is a Long Put Butterfly strategy formed, and what is the loss or profit potential of this strategy?

(b) Calculate the result of a long put butterfly strategy when selling two ATM puts and buying one ITM and one OTM put. Possible developments of the Pharma Group share range from 0.00 EUR to 140.00 EUR in steps of five. Plot the result graphically.

Content

The Long Put Butterfly strategy behaves simultaneously to the Long Call Butterfly, but this strategy is formed with puts.

The long put butterfly variant can be formed by buying one ITM put and one OTM put, while selling two ATM puts. Thus, there are three different strike prices,

Long Call Butterfly

Assumptions	
# Long Calls (OTM)	1.00
Long Call price (OTM)	6.13
Strike price of Long Call (OTM)	72.00
# Long Calls (ITM)	1.00
Long Call price (ITM)	7.67
Strike price of Long Call (ITM)	68.00
# Short Calls	2.00
Short Call price	6.86
Strike price of Short Call	70.00

Price of the underlying	Profit/loss
0.00	(0.07)
5.00	(0.07)
10.00	(0.07)
15.00	(0.07)
20.00	(0.07)
25.00	(0.07)
30.00	(0.07)
35.00	(0.07)
40.00	(0.07)
45.00	(0.07)
50.00	(0.07)
55.00	(0.07)
60.00	(0.07)
65.00	(0.07)
70.00	1.93
75.00	(0.07)
80.00	(0.07)
85.00	(0.07)
90.00	(0.07)
95.00	(0.07)
100.00	(0.07)
105.00	(0.07)
110.00	(0.07)
115.00	(0.07)
120.00	(0.07)
125.00	(0.07)
130.00	(0.07)
135.00	(0.07)
140.00	(0.07)

Received/paid premium	(0.07)

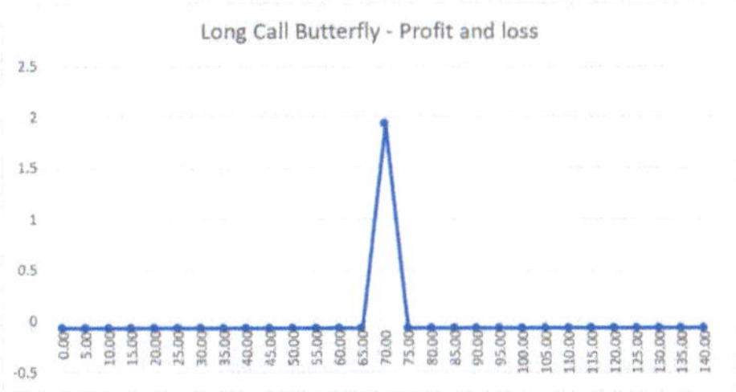

Fig. 5.7 Calculation of long call butterfly

and the maturities are identical. The profit and loss potential of this strategy are limited.

Important Formulas

$$\text{Maximum profit} = \text{strike price of the higher long put} - \text{strike price of the short put} + \text{premium received} - \text{premium paid}$$

$$\text{Higher break-even} = \text{strike price of the higher long put} + \text{premium received} - \text{premium paid}$$

$$\text{Lower break-even} = \text{strike price of the lower long put} - \text{premium received} + \text{premium paid}$$

$$\text{Maximum loss} = \text{premium received} - \text{premium paid}$$

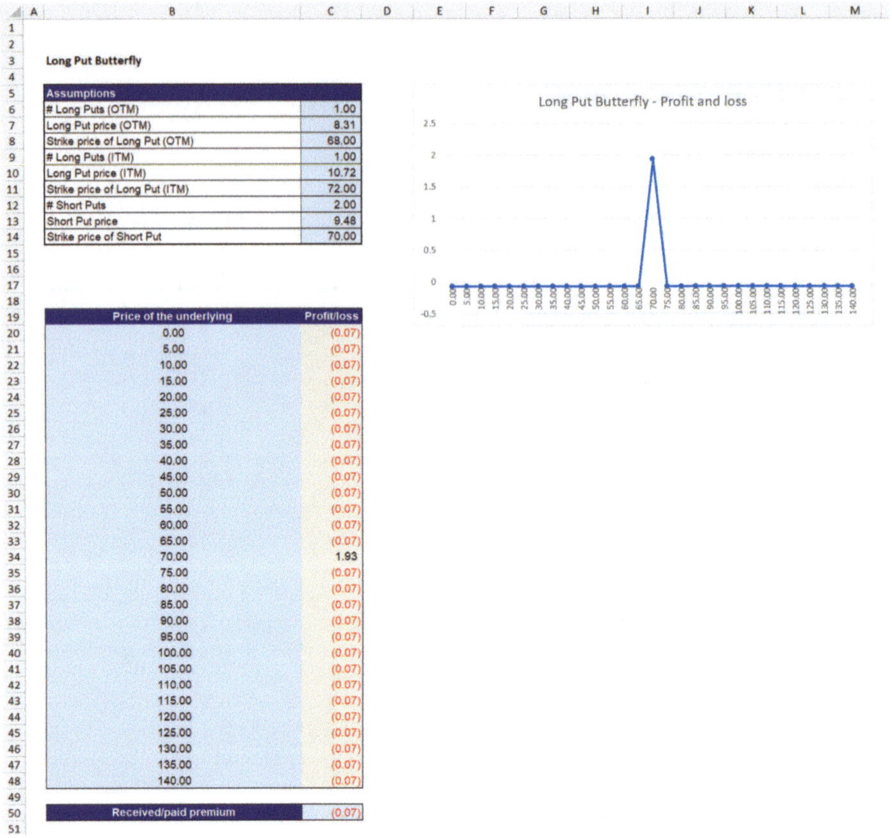

Long Put Butterfly

Assumptions	
# Long Puts (OTM)	1.00
Long Put price (OTM)	8.31
Strike price of Long Put (OTM)	68.00
# Long Puts (ITM)	1.00
Long Put price (ITM)	10.72
Strike price of Long Put (ITM)	72.00
# Short Puts	2.00
Short Put price	9.48
Strike price of Short Put	70.00

Price of the underlying	Profit/loss
0.00	(0.07)
5.00	(0.07)
10.00	(0.07)
15.00	(0.07)
20.00	(0.07)
25.00	(0.07)
30.00	(0.07)
35.00	(0.07)
40.00	(0.07)
45.00	(0.07)
50.00	(0.07)
55.00	(0.07)
60.00	(0.07)
65.00	(0.07)
70.00	1.93
75.00	(0.07)
80.00	(0.07)
85.00	(0.07)
90.00	(0.07)
95.00	(0.07)
100.00	(0.07)
105.00	(0.07)
110.00	(0.07)
115.00	(0.07)
120.00	(0.07)
125.00	(0.07)
130.00	(0.07)
135.00	(0.07)
140.00	(0.07)

Received/paid premium	(0.07)

Fig. 5.8 Calculation of long put butterfly

Display in Excel

You can enter the following formula in cell C20:

=SUM(IF(B20<$C$8;$C$8-B20-$C$7;-$C$7)*$C$6;IF(B20<$C$11;$C$11-B20-$C$10;-$C$10)*$C$9;IF(B20<$C$14;B20-$C$14+$C$13;$C$13)*$C$12)

Following this, you can drag the cell C20 to C48 with the mouse. You will then get the following picture as shown in Fig. 5.8.

5.9 Assignment 22: Neutral Bearish–Long Call Ladder

Task

(a) How is a long call ladder strategy formed, and what is the loss or profit potential of this strategy?

(b) Calculate the result of a long call ladder strategy when buying an ITM call and selling an ATM and an OTM call. Possible developments of the Pharma Group share range from 0.00 EUR to 140.00 EUR in steps of five. Plot the result graphically.

Content
The Long Call Ladder strategy is formed by the investor buying an ITM call and selling an ATM call and an OTM call at the same time. Especially in case of a strong increase in the underlying, this strategy has an unlimited loss potential. The profit potential, however, is limited.

Important Formulas

$$\text{Maximum profit} = \text{strike price of the lower short call} - \text{strike price of the long call} \\ + \text{premium received} - \text{premium paid}$$

$$\text{Higher break} - \text{even} = \text{strike price of the higher short call} + \text{maximum profit}$$

If a debit has arisen on the part of the premium, a second (lower) break-even point exists:

$$\text{Loss} = \text{higher break} - \text{even price of the underlying asset}$$

Display in Excel
You can enter the following formula in cell C20:

=SUM(IF(B20>$C$8;B20-$C$8-$C$7;-$C$7)*$C$6;IF(B20>$C$11;$C$11-B20+$C$10;$C$10)*$C$9;IF(B20>$C$14;$C$14-B20+$C$13;$C$13)*$C$12)

Following this, you can drag the cell C20 to C48 with the mouse. You will then get the following picture as shown in Fig. 5.9.

5.10 Assignment 23: Neutral Bearish–Long Put Ladder

Task

(a) How is a long put ladder strategy formed, and what is the loss or profit potential of this strategy?
(b) Calculate the result of a long put ladder strategy when buying an ITM put and selling an ATM and an OTM put. Possible developments of the Pharma Group share range from 0.00 EUR to 140.00 EUR in steps of five. Plot the result graphically.

Content
This strategy is formed by the investor buying an ITM put and selling an ATM put and an OTM put at the same time. Unlike the Long Call Ladder, the investor is not protected against a sharp decrease in the value of the underlying asset. Although the

Long Call Ladder

Assumptions	
# Long Calls	1.00
Long Call price	7.67
Strike price of Long Call	68.00
# Short Calls (OTM)	1.00
Short Call price (OTM)	6.13
Strike price of Short Call (OTM)	72.00
# Short Calls (ATM)	1.00
Short Call price (ATM)	6.86
Strike price of Short Call (ATM)	70.00

Price of the underlying	Profit/loss
0.00	5.32
5.00	5.32
10.00	5.32
15.00	5.32
20.00	5.32
25.00	5.32
30.00	5.32
35.00	5.32
40.00	5.32
45.00	5.32
50.00	5.32
55.00	5.32
60.00	5.32
65.00	5.32
70.00	7.32
75.00	4.32
80.00	(0.68)
85.00	(5.68)
90.00	(10.68)
95.00	(15.68)
100.00	(20.68)
105.00	(25.68)
110.00	(30.68)
115.00	(35.68)
120.00	(40.68)
125.00	(45.68)
130.00	(50.68)
135.00	(55.68)
140.00	(60.68)

Received/paid premium	5.32

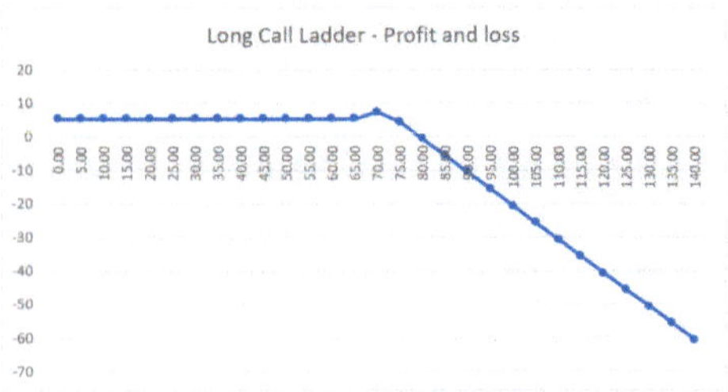

Fig. 5.9 Calculation of long call ladder

Long Put Ladder strategy is shown with an unlimited risk of loss, the underlying asset cannot fall below the value of zero. In the case of a strong increase in the value of the underlying asset, the profit with this strategy is limited.

Important Formulas

$$\text{Maximum profit} = \text{strike price of the long put} - \text{strike price of the higher short put} + \text{premium received} - \text{premium paid}$$

The maximum profit is reached when the price of the underlying is between the strike prices of the two short puts.

Display in Excel

You can enter the following formula in cell C20:

```
=SUM(IF(B20<$C$8;$C$8-B20-$C$7;-$C$7)*$C$6;IF(B20<$C$11;B20-$C$11+$C$10;$C$10)*$C$9;IF(B20<$C$14;B20-$C$14+$C$13;$C$13)*$C$12)
```

Long Put Ladder

Assumptions	
# Long Put	1.00
Long Put price	10.72
Strike price of Long Put	72.00
# Short Puts (OTM)	1.00
Short Put price (OTM)	8.31
Strike price of Short Put (OTM)	68.00
# Short Puts (ATM)	1.00
Short Put price (ATM)	9.48
Strike price of Short Put (ATM)	70.00

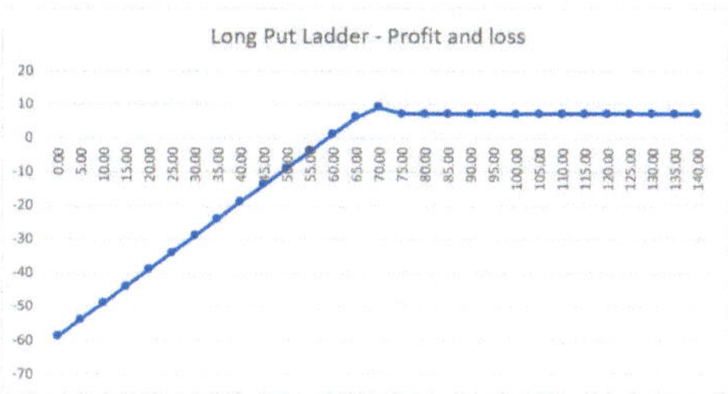

Price of the underlying	Profit/loss
0.00	(58.93)
5.00	(53.93)
10.00	(48.93)
15.00	(43.93)
20.00	(38.93)
25.00	(33.93)
30.00	(28.93)
35.00	(23.93)
40.00	(18.93)
45.00	(13.93)
50.00	(8.93)
55.00	(3.93)
60.00	1.07
65.00	6.07
70.00	9.07
75.00	7.07
80.00	7.07
85.00	7.07
90.00	7.07
95.00	7.07
100.00	7.07
105.00	7.07
110.00	7.07
115.00	7.07
120.00	7.07
125.00	7.07
130.00	7.07
135.00	7.07
140.00	7.07

Received/paid premium	7.07

Fig. 5.10 Calculation of long put ladder

Following this, you can drag the cell C20 to C48 with the mouse. You will then get the following picture as shown in Fig. 5.10.

5.11 Assignment 24: Neutral Bearish–Short Strangle

Task

(a) How is a short strangle strategy formed, and what is the loss or profit potential of this strategy?
(b) Calculate the result of a short strangle strategy when selling an OTM call and an OTM put. Possible developments of the Pharma Group share range from 0.00 EUR to 140.00 EUR in steps of five. Plot the result graphically.

Content

The short strangle is chosen if the investor assumes low volatility and does not expect the underlying to change strongly upwards or downwards. Here, an equal number of OTM calls and puts are sold, with the maturities and the underlying being identical. The investor collects the premium, from which the maximum profit of this strategy is derived.

Important Formulas

$$\text{Maximum profit} = \text{premium received}$$

$$\text{Higher break} - \text{even} = \text{strike price of the short call} + \text{premium received}$$

$$\text{Lower break} - \text{even} = \text{strike price of the short put} - \text{premium received}$$

Display in Excel

You can enter the following formula in cell C20:

=SUM(IF(B20>$C$8;$C$8-B20+$C$7;$C$7)*$C$6;IF(B20<$C$11;B20-$C$11+$C$10;$C$10)*$C$9)

Following this, you can drag the cell C20 to C48 with the mouse. You will then get the following picture as shown in Fig. 5.11.

5.12 Assignment 25: Neutral Bearish–Short Straddle

Task

(a) How is a short straddle strategy formed, and what is the loss or profit potential of this strategy?
(b) Calculate the result of a short straddle strategy when selling an ATM call and an ATM put. Possible developments of the Pharma Group share range from 0.00 EUR to 140.00 EUR in steps of five. Plot the result graphically.

Content

With the short straddle strategy, the investor also assumes a constant performance of the underlying. Since the same number of calls and puts are sold, the difference from the short straddle is that the strike price of the options is identical. Therefore, this strategy has its highest profit exactly when the price of the underlying equals the strike price, as the holder of the options does not exercise them, and the investor keeps the full premium. However, this strategy is accompanied by a high or unlimited loss potential in case of a breakout of the underlying in both directions.

Important Formulas

$$\text{Maximum profit} = \text{premium received}$$

$$\text{Higher break} - \text{even} = \text{strike price of the short call} + \text{premium received}$$

$$\text{Lower break} - \text{even} = \text{strike price of the short put} - \text{premium received}$$

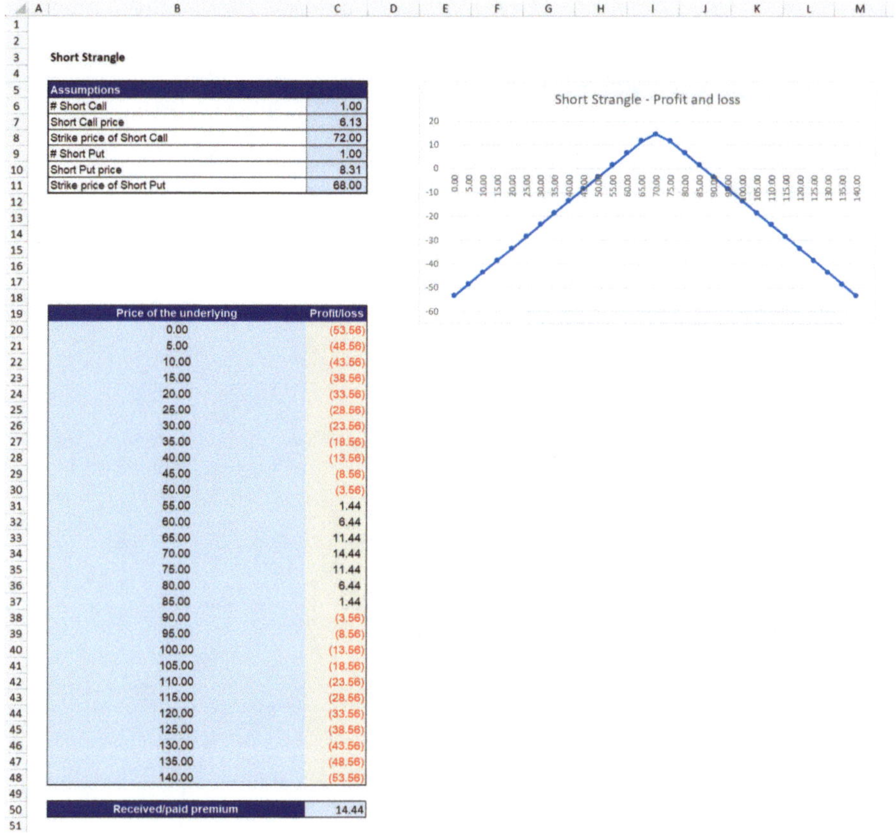

Short Strangle

Assumptions	
# Short Call	1.00
Short Call price	6.13
Strike price of Short Call	72.00
# Short Put	1.00
Short Put price	8.31
Strike price of Short Put	68.00

Price of the underlying	Profit/loss
0.00	(53.56)
5.00	(48.56)
10.00	(43.56)
15.00	(38.56)
20.00	(33.56)
25.00	(28.56)
30.00	(23.56)
35.00	(18.56)
40.00	(13.56)
45.00	(8.56)
50.00	(3.56)
55.00	1.44
60.00	6.44
65.00	11.44
70.00	14.44
75.00	11.44
80.00	6.44
85.00	1.44
90.00	(3.56)
95.00	(8.56)
100.00	(13.56)
105.00	(18.56)
110.00	(23.56)
115.00	(28.56)
120.00	(33.56)
125.00	(38.56)
130.00	(43.56)
135.00	(48.56)
140.00	(53.56)

Received/paid premium	14.44

Fig. 5.11 Calculation of short strangle

$$\text{Loss} = \text{strike price of the short call} + \text{price of the underlying} + \text{premium received or price of the underlying} - \text{strike price of the short put} + \text{premium received}$$

Display in Excel

You can enter the following formula in cell C20:

=SUM(IF(B20>$C$8;$C$8-B20+$C$7;$C$7)*$C$6;IF(B20<$C$11;B20-$C$11+$C$10;$C$10)*$C$9)

Following this, you can drag the cell C20 to C48 with the mouse. You will then get the following picture as shown in Fig. 5.12.

Short Straddle

Assumptions	
# Short Call	1.00
Short Call price	6.86
Strike price of Short Call	70.00
# Short Put	1.00
Short Put price	9.48
Strike price of Short Put	70.00

Price of the underlying	Profit/loss
0.00	(53.65)
5.00	(48.65)
10.00	(43.65)
15.00	(38.65)
20.00	(33.65)
25.00	(28.65)
30.00	(23.65)
35.00	(18.65)
40.00	(13.65)
45.00	(8.65)
50.00	(3.65)
55.00	1.35
60.00	6.35
65.00	11.35
70.00	16.35
75.00	11.35
80.00	6.35
85.00	1.35
90.00	(3.65)
95.00	(8.65)
100.00	(13.65)
105.00	(18.65)
110.00	(23.65)
115.00	(28.65)
120.00	(33.65)
125.00	(38.65)
130.00	(43.65)
135.00	(48.65)
140.00	(53.65)

Received/paid premium	16.35

Fig. 5.12 Calculation of short straddle

5.13 Assignment 26: Neutral Bearish–Short Guts

Task

(a) How is a Short Guts strategy formed, and what is the loss or profit potential of this strategy?
(b) Calculate the result of a short guts strategy when selling an ITM call and an ITM put. Possible developments of the Pharma Group share range from 0.00 EUR to 140.00 EUR in steps of five. Plot the result graphically.

Content

The Short Guts strategy is identical to the Short Strangle strategy, with the difference that the investor in this strategy does not sell OTM calls and puts, but ITM calls and puts. As a result, the premium collected is correspondingly higher. The profit potential of this strategy is limited. The loss potential is very high or unlimited.

Important Formulas

$$\text{Maximum profit} = \text{premium received}$$

$$\text{Higher break} - \text{even} = \text{strike price of the short call} + \text{premium received}$$

$$\text{Lower break} - \text{even} = \text{strike price of the short put} - \text{premium received}$$

$$\begin{aligned}\text{Loss} &= \text{strike price of the short call} + \text{price of the underlying}\\ &\quad + \text{premium received or price of the underlying} - \text{strike price of the short put}\\ &\quad + \text{premium received}\end{aligned}$$

Display in Excel

You can enter the following formula in cell C20:

=SUM(IF(B20>$C$8;$C$8-B20+$C$7;$C$7)*$C$6;IF(B20<$C$11;B20-$C$11+$C$10;$C$10)*$C$9)

Following this, you can drag the cell C20 to C48 with the mouse. You will then get the following picture as shown in Fig. 5.13.

5.14 Assignment 27: Neutral Bullish–Short Condor

Task

(a) How is a Short Condor strategy formed, and what is the loss or profit potential of this strategy?
(b) Calculate the result of a Short Condor strategy when buying an OTM call, an ITM call, and selling an ITM call (lower strike price) and an OTM call (higher strike price). The data for this can be found in the previous disclosures. Possible developments of the Pharma Group share range from 0.00 EUR to 140.00 EUR in steps of five. Plot the result graphically.

Content

The Short Condor is the counterpart to the Long Condor or Condor Options. The investor usually receives a positive premium result with the Short Condor; therefore, the Short Condor is a credit spread strategy. It is formed by buying an ITM and an OTM call and selling an ITM call with a lower strike price and an OTM call with a higher strike price. The profit and loss potential of this strategy are limited.

Important Formulas

$$\text{Maximum profit} = \text{premium received} - \text{premium paid}$$

Due to its construction, the Short Condor strategy also usually has two break-even points:

$$\begin{aligned}\text{Higher break} - \text{even} &= \text{strike price of the higher short call}\\ &\quad - \text{premium received} + \text{premium paid}\end{aligned}$$

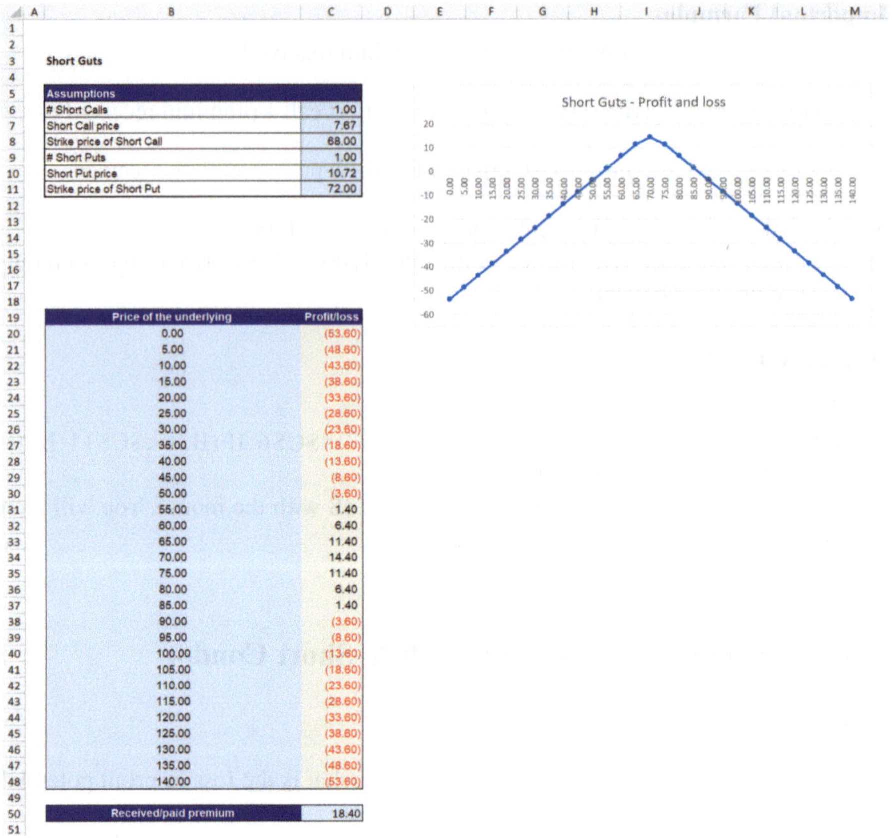

Short Guts

Assumptions	
# Short Calls	1.00
Short Call price	7.67
Strike price of Short Call	68.00
# Short Puts	1.00
Short Put price	10.72
Strike price of Short Put	72.00

Price of the underlying	Profit/loss
0.00	(53.60)
5.00	(48.60)
10.00	(43.60)
15.00	(38.60)
20.00	(33.60)
25.00	(28.60)
30.00	(23.60)
35.00	(18.60)
40.00	(13.60)
45.00	(8.60)
50.00	(3.60)
55.00	1.40
60.00	6.40
65.00	11.40
70.00	14.40
75.00	11.40
80.00	6.40
85.00	1.40
90.00	(3.60)
95.00	(8.60)
100.00	(13.60)
105.00	(18.60)
110.00	(23.60)
115.00	(28.60)
120.00	(33.60)
125.00	(38.60)
130.00	(43.60)
135.00	(48.60)
140.00	(53.60)

Received/paid premium	18.40

Fig. 5.13 Calculation of Short Guts

$$\text{Lower break} - \text{even} = \text{strike price lower short call} + \text{premium received} - \text{premium paid}$$

$$\text{Maximum loss} = \text{strike price of lower short call} - \text{strike price of lower long call} + \text{premium received} - \text{premium paid}$$

Display in Excel

You can enter the following formula in cell C20:

=SUM(IF(B20>$C$8;B20-$C$8-$C$7;-$C$7)*$C$6;IF(B20>$C$11;B20-$C$11-$C$10;-$C$10)*$C$9;IF(B20>$C$14;$C$14-B20+$C$13;$C$13)*$C$12;IF(B20>$C$17;$C$17-B20+$C$16;$C$16)*$C$15)

Following this, you can drag the cell C20 to C48 with the mouse. You will then get the following picture as shown in Fig. 5.14.

Short Condor Strategy

Assumptions	
# Long Call (OTM)	1.00
Long Call price (OTM)	6.13
Strike price of Long Call (OTM)	72.00
# Long Call (ITM)	1.00
Long Call price (ITM)	7.67
Strike price of Long Call (ITM)	68.00
# Short Calls (OTM higher strike price)	1.00
Short Call price (OTM higher strike price)	5.46
Strike price of Short Call (OTM higher strike price)	74.00
# Short Calls (ITM lower strike price)	1.00
Short Call price (ITM lower strike price)	8.56
Strike price of Short Call (ITM lower strike price)	66.00

Price of the underlying	Profit/loss
0.00	0.22
5.00	0.22
10.00	0.22
15.00	0.22
20.00	0.22
25.00	0.22
30.00	0.22
35.00	0.22
40.00	0.22
45.00	0.22
50.00	0.22
55.00	0.22
60.00	0.22
65.00	0.22
70.00	(1.78)
75.00	0.22
80.00	0.22
85.00	0.22
90.00	0.22
95.00	0.22
100.00	0.22
105.00	0.22
110.00	0.22
115.00	0.22
120.00	0.22
125.00	0.22
130.00	0.22
135.00	0.22
140.00	0.22

Received/paid premium	0.22

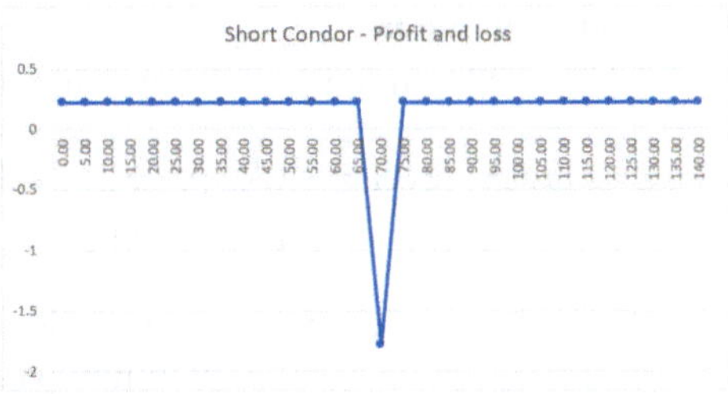

Fig. 5.14 Calculation of Short Condor

5.15 Assignment 28: Neutral Bullish–Short Call Butterfly

Task

(a) How is a Short Call Butterfly strategy formed, and what is the loss or profit potential of this strategy?

(b) Calculate the result of a short call butterfly strategy in which one ITM and one OTM call are sold, whereas two ATM calls are bought. Possible developments of the Pharma Group share range from 0.00 EUR to 140.00 EUR in steps of five. Plot the result graphically.

Content

The short call butterfly is a strategy in which the investor assumes an upward or downward movement of the underlying. It is not suitable if a constant development is expected. In a short call butterfly, one ITM call and one OTM call are sold, whereas two ATM calls are bought. The maturities and the underlying are identical.

In this construction, the Short Call Butterfly is usually a credit spread strategy. The profit and loss potential of this strategy are limited.

Important Formulas

$$\text{Maximum profit} = \text{premium received} - \text{premium paid}$$

$$\text{Higher break} - \text{even} = \text{strike price of the higher short call} - \text{premium received} + \text{premium paid}$$

$$\text{Lower break} - \text{even} = \text{strike price lower short call} + \text{premium received} - \text{premium paid}$$

$$\text{Maximum loss} = \text{strike price of lower short call} - \text{strike price of long call} + \text{premium received} - \text{premium paid}$$

Display in Excel

You can enter the following formula in cell C20:

=SUM(IF(B20>$C$8;B20-$C$8-$C$7;-$C$7)*$C$6;IF(B20>$C$11;$C$11-B20+$C$10;$C$10)*$C$9;IF(B20>$C$14;$C$14-B20+$C$13;$C$13)*$C$12)

Following this, you can drag the cell C20 to C48 with the mouse. You will then get the following picture as shown in Fig. 5.15.

5.16 Assignment 29: Neutral Bullish–Short Put Butterfly

Task

(a) How is a Short Put Butterfly strategy formed, and what is the loss or profit potential of this strategy?
(b) Calculate the result of a short put butterfly strategy in which one ITM and one OTM put are sold, whereas two ATM puts are bought. Possible developments of the Pharma Group share range from 0.00 EUR to 140.00 EUR in steps of five. Plot the result graphically.

Content

The short put butterfly is similar to the short call butterfly. Here, too, the investor assumes a downward or upward breakout of the underlying. Only in the case of a constant development does the investor experience a loss. However, the short put butterfly is formed with put options. For this purpose, an investor can buy two ATM puts and sell one ITM put and one OTM put at the same time. The strike prices of the ITM and OTM puts should be chosen in such a way that the investor receives a premium surplus. This credit is also the highest possible profit from this strategy. The profit and loss potential of this strategy are limited.

Short Call Butterfly

Assumptions	
# Long Call	2.00
Long Call price	6.86
Strike price of Long Call	70.00
# Short Calls (OTM)	1.00
Short Call price (OTM)	6.13
Strike price of Short Call (OTM)	72.00
# Short Calls (ITM)	1.00
Short Call price (ITM)	7.67
Strike price of Short Call (ITM)	68.00

Price of the underlying value	Profit/loss
0.00	0.07
5.00	0.07
10.00	0.07
15.00	0.07
20.00	0.07
25.00	0.07
30.00	0.07
35.00	0.07
40.00	0.07
45.00	0.07
50.00	0.07
55.00	0.07
60.00	0.07
65.00	0.07
70.00	(1.93)
75.00	0.07
80.00	0.07
85.00	0.07
90.00	0.07
95.00	0.07
100.00	0.07
105.00	0.07
110.00	0.07
115.00	0.07
120.00	0.07
125.00	0.07
130.00	0.07
135.00	0.07
140.00	0.07
Received/paid premium	0.07

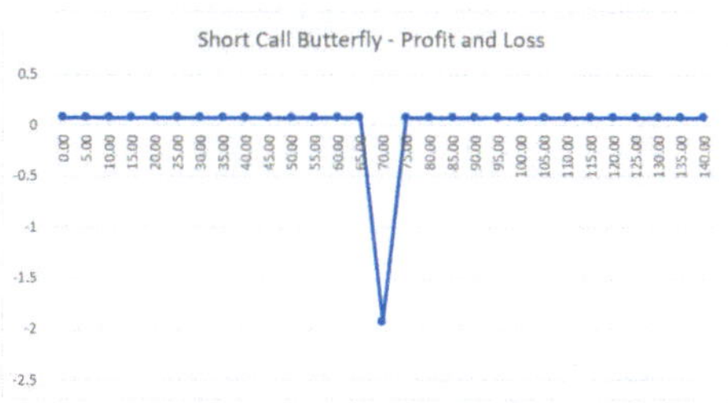

Fig. 5.15 Calculation of short call butterfly

Important Formulas

$$\text{Maximum profit} = \text{premium received} - \text{premium paid}$$

$$\text{Higher break-even} = \text{strike price of the higher short put} - \text{premium received} + \text{premium paid}$$

$$\text{Lower break-even} = \text{strike price lower short put} + \text{premium received} - \text{premium paid}$$

$$\text{Maximum loss} = \text{strike price of the lower short put} - \text{strike price of the long put} + \text{premium received} - \text{premium paid}$$

Display in Excel

You can enter the following formula in cell C20:

```
=SUM(IF(B20<$C$8;$C$8-B20-$C$7;-$C$7)*$C$6;IF(B20<$C$11;B20-$C$11+$C$10;$C$10)*$C$9;IF(B20<$C$14;B20-$C$14+$C$13;$C$13)*$C$12)
```

Short Put Butterfly

Assumptions	
# Long Put	2.00
Long Put price	9.48
Strike price of Long Put	70.00
# Short Puts (OTM)	1.00
Short Put price (OTM)	8.31
Strike price of Short Put (OTM)	68.00
# Short Puts (ITM)	1.00
Short Put price (ITM)	10.72
Strike Price of Short Put (ITM)	72.00

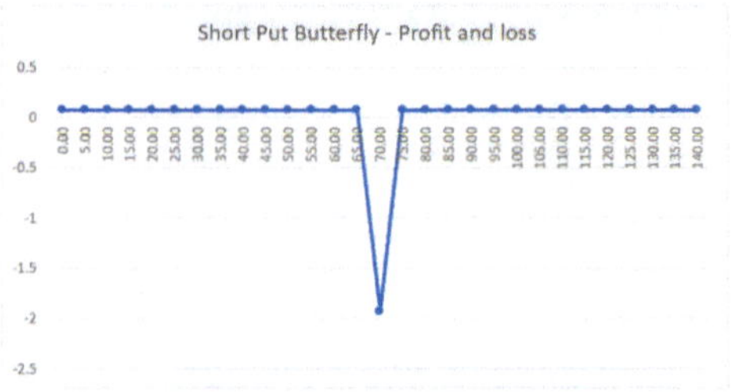

Price of the underlying	Profit/loss
0.00	0.07
5.00	0.07
10.00	0.07
15.00	0.07
20.00	0.07
25.00	0.07
30.00	0.07
35.00	0.07
40.00	0.07
45.00	0.07
50.00	0.07
55.00	0.07
60.00	0.07
65.00	0.07
70.00	(1.93)
75.00	0.07
80.00	0.07
85.00	0.07
90.00	0.07
95.00	0.07
100.00	0.07
105.00	0.07
110.00	0.07
115.00	0.07
120.00	0.07
125.00	0.07
130.00	0.07
135.00	0.07
140.00	0.07

Received/paid premium	0.07

Fig. 5.16 Calculation of short put butterfly

Following this, you can drag the cell C20 to C48 with the mouse. You will then get the following picture as shown in Fig. 5.16.

5.17 Assignment 30: Neutral Bullish–Short Call Ladder

Task

(a) How is a short call ladder strategy formed, and what is the loss or profit potential of this strategy?

(b) Calculate the result of a short call ladder strategy in which an ATM and an OTM call are bought, and an ITM call is sold. Possible developments of the Pharma Group share range from 0.00 EUR to 140.00 EUR in steps of five. Plot the result graphically.

Content

In the Short Call Ladder strategy, the investor assumes that the underlying will not experience constant development. He expects an upward breakout, in which case his profit potential is unlimited. At the same time, he also experiences limited profit in the event of a downward development of the underlying. The construction of the short call ladder provides for the sale of an ITM call and the purchase of an ATM and an OTM call. The maturities and the underlying are identical. The profit potential of this strategy is unlimited as the underlying rises. The loss potential is limited as the underlying cannot be less than zero.

Important Formulas

$$\text{Profit} = \text{price of the underlying asset} - \text{higher break} - \text{even}$$

$$\text{Higher break} - \text{even} = \text{strike price of the higher long call} + \text{maximum loss}$$

If the premium of the bought and sold options balances out or a credit for the investor arises, this strategy receives a second break-even.

$$\begin{aligned}\text{Lower break} - \text{even} = {} & \text{strike price of the short call} - \text{premium received} \\ & + \text{premium paid}\end{aligned}$$

$$\begin{aligned}\text{Maximum loss} = {} & \text{strike price of lower short call} - \text{strike price of lower long call} \\ & + \text{premium received} - \text{premium paid}\end{aligned}$$

Display in Excel

You can enter the following formula in cell C20:

```
=SUM(IF(B20>$C$8;B20-$C$8-$C$7;-$C$7)*$C$6;IF(B20>$C$11;B20-$C$11-
   $C$10;-$C$10)*$C$9;IF(B20>$C$14;$C$14-B20+$C$13;$C$13)*$C$12)
```

Following this, you can drag the cell C20 to C48 with the mouse. You will then get the following picture as shown in Fig. 5.17.

5.18 Assignment 31: Neutral Bullish–Short Put Ladder

Task

(a) How is a short put ladder strategy formed, and what is the loss or profit potential of this strategy?
(b) Calculate the result of a short put ladder strategy where an ATM and an OTM put are bought, and an ITM put is sold. Possible developments of the Pharma Group share range from 0.00 EUR to 140.00 EUR in steps of five. Plot the result graphically.

Content

The short put ladder is the counterpart to the short call ladder. Here, the investor bets that the underlying will experience a strong downward movement, as his profit potential is almost unlimited, at least up to the natural limit when the underlying

Short Call Ladder

Assumptions	
# Long Calls (OTM)	1.00
Long Call price (OTM)	6.13
Strike price of Long Call (OTM)	72.00
# Long Calls (ATM)	1.00
Long Call price (ATM)	6.86
Strike price of Long Call (ATM)	70.00
# Short Calls	1.00
Short Call price	7.67
Strike price of Short Call	68.00

Price of the underlying	Profit/loss
0.00	(5.32)
5.00	(5.32)
10.00	(5.32)
15.00	(5.32)
20.00	(5.32)
25.00	(5.32)
30.00	(5.32)
35.00	(5.32)
40.00	(5.32)
45.00	(5.32)
50.00	(5.32)
55.00	(5.32)
60.00	(5.32)
65.00	(5.32)
70.00	(7.32)
75.00	(4.32)
80.00	0.68
85.00	5.68
90.00	10.68
95.00	15.68
100.00	20.68
105.00	25.68
110.00	30.68
115.00	35.68
120.00	40.68
125.00	45.68
130.00	50.68
135.00	55.68
140.00	60.68

Received/paid premium	(5.32)

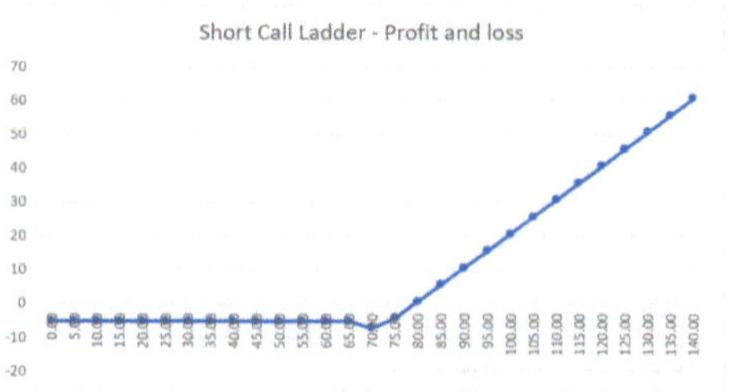

Fig. 5.17 Calculation of short call ladder

takes the value zero. In the case of an upward movement of the underlying, its profit or loss is limited, depending on whether the investor achieves a credit or a debit after offsetting the premiums. In the case of a constant movement of the underlying, the investor experiences his highest loss. The construction of a short put ladder consists of the sale of an ITM put and the purchase of an ATM and an OTM put.

Important Formulas

$$\text{Profit} = \text{lower break} - \text{even} - \text{price of the underlying asset}$$

$$\text{Lower break} - \text{even} = \text{strike price of the lower long put} + \text{maximum loss}$$

If the premium of the bought and sold options balances out or a credit for the investor arises, this strategy receives a second break-even.

$$\text{Higher break} - \text{even} = \text{strike price of the short put} - \text{premium received} + \text{premium paid}$$

$$\text{Maximum loss} = \text{strike price of the higher long put} - \text{strike price of the short put} + \text{premium received} - \text{premium paid}$$

Short Put Ladder

Assumptions	
# Long Puts (OTM)	1.00
Long Put price (OTM)	8.31
Strike price of Long Put (OTM)	68.00
# Long Puts (ATM)	1.00
Long Put price (ATM)	9.48
Strike price of Long Put (ATM)	70.00
# Short Puts	1.00
Short Put price	10.72
Strike price of Short Put	72.00

Price of the underlying	Profit/loss
0.00	58.93
5.00	53.93
10.00	48.93
15.00	43.93
20.00	38.93
25.00	33.93
30.00	28.93
35.00	23.93
40.00	18.93
45.00	13.93
50.00	8.93
55.00	3.93
60.00	(1.07)
65.00	(6.07)
70.00	(9.07)
75.00	(7.07)
80.00	(7.07)
85.00	(7.07)
90.00	(7.07)
95.00	(7.07)
100.00	(7.07)
105.00	(7.07)
110.00	(7.07)
115.00	(7.07)
120.00	(7.07)
125.00	(7.07)
130.00	(7.07)
135.00	(7.07)
140.00	(7.07)

Received/paid premium	(7.07)

Fig. 5.18 Calculation of short put ladder

Display in Excel

You can enter the following formula in cell C20:

=SUM(IF(B20<$C$8;$C$8-B20-$C$7;-$C$7)*$C$6;IF(B20<$C$11;$C$11-B20-$C$10;-$C$10)*$C$9;IF(B20<$C$14;B20-$C$14+$C$13;$C$13)*$C$12)

Following this, you can drag the cell C17 to C45 with the mouse. You will then get the following picture as shown in Fig. 5.18.

5.19 Assignment 32: Neutral Bullish–Long Strangle

Task

(a) How is a long strangle strategy formed, and what is the loss or profit potential of this strategy?

(b) Calculate the result of a long strangle strategy in which an OTM call and an OTM put are bought. Possible developments of the Pharma Group share range from 0.00 EUR to 140.00 EUR in steps of five. Show the result graphically.

Content

For the investor in a long strangle, it is not important whether the underlying is experiencing a strong upward or downward movement. His profit potential is unlimited in both directions, while the loss potential is limited. In case of a constant or only moderate movement, however, it experiences a loss. The long strangle consists of buying the same number of OTM calls and puts.

Important Formulas

$$\begin{aligned}\text{Profit} &= \text{price of the underlying} - \text{strike price of the long call}\\ &\quad - \text{paid premium or strike price of the long put} - \text{price of the underlying}\\ &\quad - \text{paid premium}\end{aligned}$$

$$\text{Higher break} - \text{even} = \text{strike price of the long call} + \text{premium paid}$$

$$\text{Lower break} - \text{even} = \text{strike price of the long put} - \text{premium paid}$$

$$\text{Maximum loss} = \text{premium paid}$$

Display in Excel

You can enter the following formula in cell C20:

=SUM(IF(B20>$C$8;B20-$C$8-$C$7;-$C$7)*$C$6;IF(B20<$C$11;$C$11-B20-$C$10;-$C$10)*$C$9)

Following this, you can drag the cell C20 to C48 with the mouse. You will then get the following picture as shown in Fig. 5.19.

5.20 Assignment 33: Neutral Bullish–Long Straddle

Task

(a) How is a long straddle strategy formed, and what is the loss or profit potential of this strategy?
(b) Calculate the result of a long straddle strategy in which an ATM put and an ATM call are bought. Possible developments of the Pharma Group share range from 0.00 EUR to 140.00 EUR in steps of five. Show the result graphically.

Content

As with the long strangle, the profit potential of a long straddle is unlimited in both directions. However, this strategy is a bit more conservative because even with smaller breakouts, the investor comes into the profit zone. Since the long straddle

Long Strangle

Assumptions	
# Long Calls	1.00
Long Call price	6.13
Strike price of Long Call	72.00
# Long Puts	1.00
Long Put price	8.31
Strike price of Long Put	68.00

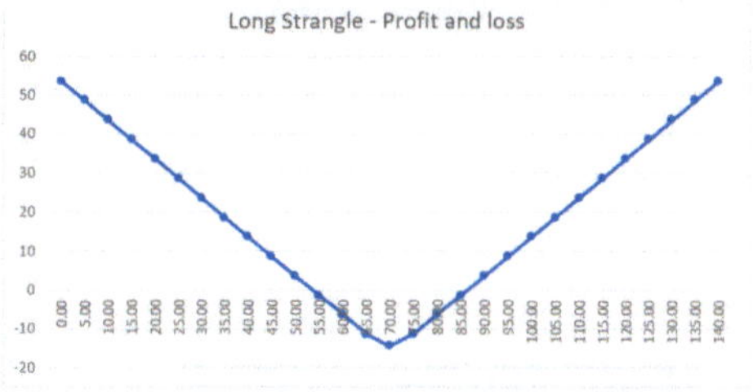

Price of the underlying	Profit/loss
0.00	53.56
5.00	48.56
10.00	43.56
15.00	38.56
20.00	33.56
25.00	28.56
30.00	23.56
35.00	18.56
40.00	13.56
45.00	8.56
50.00	3.56
55.00	(1.44)
60.00	(6.44)
65.00	(11.44)
70.00	(14.44)
75.00	(11.44)
80.00	(6.44)
85.00	(1.44)
90.00	3.56
95.00	8.56
100.00	13.56
105.00	18.56
110.00	23.56
115.00	28.56
120.00	33.56
125.00	38.56
130.00	43.56
135.00	48.56
140.00	53.56

Received/paid premium	(14.44)

Fig. 5.19 Calculation of long strangle

consists of buying the same number of calls and puts with the identical underlying, the point of maximum loss for the investor is when the underlying equals the strike price of the options (which is higher than in the straddle).

Important Formulas

$$\begin{aligned}\text{Profit} = {} & \text{price of the underlying} - \text{strike price of the long call} \\ & - \text{paid premium or strike price of the long put} - \text{price of the underlying} \\ & - \text{paid premium}\end{aligned}$$

$$\text{Higher break} - \text{even} = \text{strike price long call} + \text{paid premium}$$

$$\text{Lower break} - \text{even} = \text{strike price long put} - \text{paid premium}$$

$$\text{Maximum loss} = \text{premium paid}$$

Display in Excel

You can enter the following formula in cell C20:

Long Straddle

Assumptions	
# Long Calls	1.00
Long Call price	6.86
Strike price of Long Call	70.00
# Long Puts	1.00
Long Put price	9.48
Strike price of Long Put	70.00

Price of the underlying	Profit/loss
0.00	53.65
5.00	48.65
10.00	43.65
15.00	38.65
20.00	33.65
25.00	28.65
30.00	23.65
35.00	18.65
40.00	13.65
45.00	8.65
50.00	3.65
55.00	(1.35)
60.00	(6.35)
65.00	(11.35)
70.00	(16.35)
75.00	(11.35)
80.00	(6.35)
85.00	(1.35)
90.00	3.65
95.00	8.65
100.00	13.65
105.00	18.65
110.00	23.65
115.00	28.65
120.00	33.65
125.00	38.65
130.00	43.65
135.00	48.65
140.00	53.65

Received/paid premium	(16.35)

Fig. 5.20 Calculation of long straddle

=SUM(IF(B20>$C$8;B20-$C$8-$C$7;-$C$7)*$C$6;IF(B20<$C$11;$C$11-B20-$C$10;-$C$10)*$C$9)

Following this, you can drag the cell C20 to C48 with the mouse. You will then get the following picture as shown in Fig. 5.20.

5.21 Assignment 34: Neutral Bullish–Strip

Task

(a) How is a strip strategy formed, and what is the loss or profit potential of this strategy?

(b) Calculate the result of a strip strategy where two ATM puts and one ATM call are bought. Possible developments of the Pharma Group share range from 0.00 EUR to 140.00 EUR in steps of five. Show the result graphically.

Strip

Assumptions	
# Long Calls	1.00
Long Call price	6.86
Strike Price of Long Call	70.00
# Long Puts	2.00
Long Put price	9.48
Strike price of Long Put	70.00

Price of the underlying	Profit/loss
0.00	114.17
5.00	104.17
10.00	94.17
15.00	84.17
20.00	74.17
25.00	64.17
30.00	54.17
35.00	44.17
40.00	34.17
45.00	24.17
50.00	14.17
55.00	4.17
60.00	(5.83)
65.00	(15.83)
70.00	(25.83)
75.00	(20.83)
80.00	(15.83)
85.00	(10.83)
90.00	(5.83)
95.00	(0.83)
100.00	4.17
105.00	9.17
110.00	14.17
115.00	19.17
120.00	24.17
125.00	29.17
130.00	34.17
135.00	39.17
140.00	44.17

Received/paid premium	(25.83)

Fig. 5.21 Calculation of strip

Content

The strip is a modified version of the straddle. In this strategy, the investor has a more bearish stance and, therefore, buys more puts than calls. The strike prices, the maturities, and the underlying of the options are identical. The profit potential is unlimited, and the loss potential of this strategy is limited.

Important Formulas

$$\text{Profit} = \text{price of the underlying asset} - \text{strike price of the long call} \\ - \text{paid premium or number of puts}^{*} \\ (\text{strike price of the long put} - \text{price of the underlying asset}) - \text{paid premium}$$

$$\text{Higher break-even} = \text{base price of options} + \text{premium paid}$$

$$\text{Lower break-even} = \text{base price of options} - (\text{premium paid}/2)$$

$$\text{Maximum loss} = \text{premium paid}$$

Display in Excel
You can enter the following formula in cell C20:

=SUM(IF(B20>$C$8;B20-$C$8-$C$7;-$C$7)*$C$6;IF(B20<$C$11;$C$11-B20-$C$10;-$C$10)*$C$9)

Following this, you can drag the cell C20 to C48 with the mouse. You will then get the following picture as shown in Fig. 5.21.

5.22 Assignment 35: Neutral Bullish–Strap

Task

(a) How is a strategy strap formed, and what is the loss or profit potential of this strategy?
(b) Calculate the result of a Strap Strategy where one ATM Put and two ATM Calls were bought. Possible developments of the Pharma Group share range from 0.00 EUR to 140.00 EUR in steps of five. Show the result graphically.

Content
Like the strip, the strap is a modified version of the straddle. In this strategy, the investor now has a more bullish stance and therefore buys more calls than puts. The strike price, maturities, and the underlying are identical. The profit potential of this strategy is unlimited, and the risk of loss is limited.

Important Formulas

$$\begin{aligned}\text{Profit} = {} & \text{number of calls}^{*}\left(\text{price of the underlying asset} - \text{strike price of the long call}\right)\\ & - \text{paid premium or strike price of the long put} - \text{price of the underlying asset}\\ & - \text{paid premium}\end{aligned}$$

$$\text{Higher break} - \text{even} = \text{base price of options} + \left(\text{premium paid} / 2\right)$$

$$\text{Lower break} - \text{even} = \text{base price of options} - \text{paid premium}$$

$$\text{Maximum loss} = \text{premium paid}$$

Display in Excel
You can enter the following formula in cell C20:

=SUM(IF(B20>$C$8;B20-$C$8-$C$7;-$C$7)*$C$6;IF(B20<$C$11;$C$11-B20-$C$10;-$C$10)*$C$9)

Following this, you can drag the cell C20 to C48 with the mouse. You will then get the following picture as shown in Fig. 5.22.

Strap

Assumptions	
# Long Calls	2.00
Long Call price	6.86
Strike price of Long Call	70.00
# Long Puts	1.00
Long Put price	9.48
Strike price of Long Put	70.00

Price of the underlying	Profit/loss
0.00	46.79
5.00	41.79
10.00	36.79
15.00	31.79
20.00	26.79
25.00	21.79
30.00	16.79
35.00	11.79
40.00	6.79
45.00	1.79
50.00	(3.21)
55.00	(8.21)
60.00	(13.21)
65.00	(18.21)
70.00	(23.21)
75.00	(13.21)
80.00	(3.21)
85.00	6.79
90.00	16.79
95.00	26.79
100.00	36.79
105.00	46.79
110.00	56.79
115.00	66.79
120.00	76.79
125.00	86.79
130.00	96.79
135.00	106.79
140.00	116.79

Received/paid premium	(23.21)

Fig. 5.22 Strap calculation

5.23 Assignment 36: Neutral Bullish–Long Guts

Task

(a) How is a Long Guts strategy formed, and what is the loss or profit potential of this strategy?
(b) Calculate the result of a long guts strategy in which an ITM put and an ITM call were bought. Possible developments of the Pharma Group share range from 0.00 EUR to 140.00 EUR in steps of five. Show the result graphically.

Content

Long guts, like the long strangle, is formed by buying the same number of puts and calls with different strike prices. However, ITM options are used to construct a long guts. If the strike prices are symmetrically swapped, as in this case, the results are the same in both cases. The profit potential of this strategy is unlimited, and the risk of loss is limited.

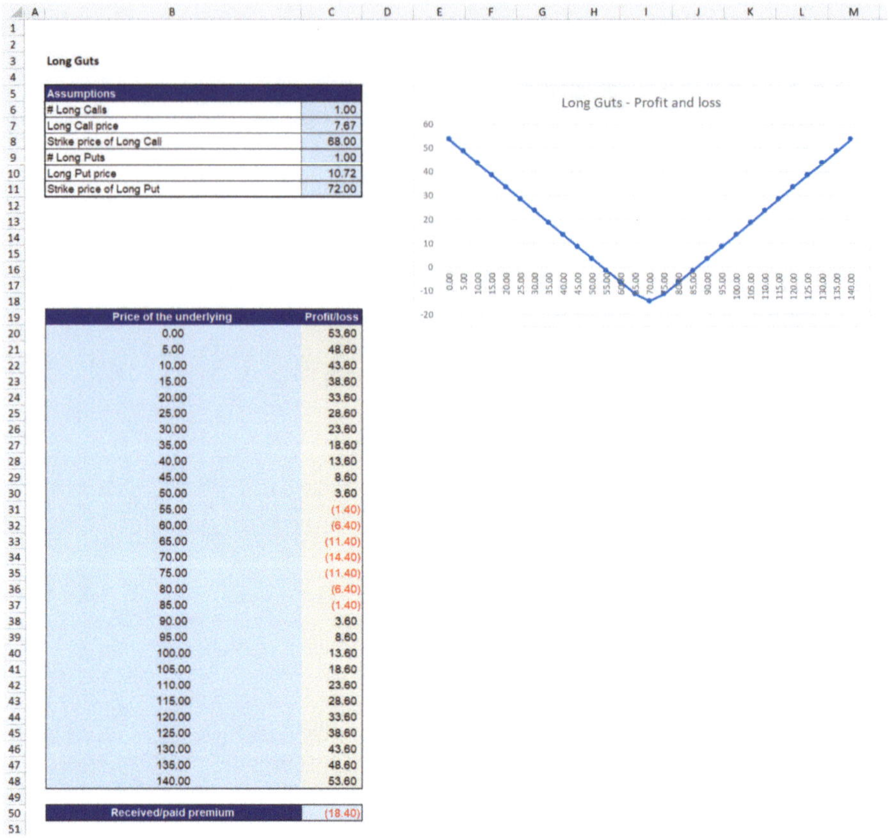

Assumptions	
# Long Calls	1.00
Long Call price	7.67
Strike price of Long Call	68.00
# Long Puts	1.00
Long Put price	10.72
Strike price of Long Put	72.00

Price of the underlying	Profit/loss
0.00	53.60
5.00	48.60
10.00	43.60
15.00	38.60
20.00	33.60
25.00	28.60
30.00	23.60
35.00	18.60
40.00	13.60
45.00	8.60
50.00	3.60
55.00	(1.40)
60.00	(6.40)
65.00	(11.40)
70.00	(14.40)
75.00	(11.40)
80.00	(6.40)
85.00	(1.40)
90.00	3.60
95.00	8.60
100.00	13.60
105.00	18.60
110.00	23.60
115.00	28.60
120.00	33.60
125.00	38.60
130.00	43.60
135.00	48.60
140.00	53.60

Received/paid premium	(18.40)

Fig. 5.23 Calculation of long guts

Important Formulas

$$\text{Profit} = \text{price of the underlying} - \text{strike price of the long call} - \text{paid premium or strike price of the long put} - \text{price of the underlying} - \text{paid premium}$$

$$\text{Higher break-even} = \text{strike price of the long call} + \text{premium paid}$$

$$\text{Lower break-even} = \text{strike price of the long put} - \text{premium paid}$$

$$\text{Maximum loss} = \text{premium paid}$$

Display in Excel

You can enter the following formula in cell C20:

```
=SUM(IF(B20>$C$8;B20-$C$8-$C$7;-$C$7)*$C$6;IF(B20<$C$11;
$C$11-B20-$C$10;-$C$10)*$C$9)
```

Following this, you can drag the cell C20 to C48 with the mouse. You will then get the following picture as shown in Fig. 5.23.

Further Reading

See Excel file Case Study Derivatives Part 2, Excel worksheet 20.
See Excel file Case Study Derivatives Part 2, Excel worksheet 21.
See video "Put Backspread, Bear Put Spread, Bear Call Spread."
See Excel file Case Study Derivatives Part 2, Excel worksheet 22.
See video "Put Backspread, Bear Put Spread, Bear Call Spread."
See Excel file Case Study Derivatives Part 2, Excel worksheet 23.
See video "Put Backspread, Bear Put Spread, Bear Call Spread."
See Excel file Case Study Derivatives Part 2, Excel worksheet 24.
See Excel file Case Study Derivatives Part 2, Excel worksheet 25.
See Excel file Case Study Derivatives Part 2, Excel worksheet 26.
See video "Butterfly."
See Excel file Case Study Derivatives Part 2, Excel worksheet 27.
See video "Butterfly."
See Excel file Case Study Derivatives Part 2, Excel worksheet 28.
See Excel file Case Study Derivatives Part 2, Excel worksheet 29.
See Excel file Case Study Derivatives Part 2, Excel worksheet 30.
See video "Straddle and Strangle."
See Excel file Case Study Derivatives Part 2, Excel worksheet 31.
See video "Straddle and Strangle."
See Excel file Case Study Derivatives Part 2, Excel worksheet 32.
See Excel file Case Study Derivatives Part 2, Excel worksheet 33.
See Excel file Case Study Derivatives Part 2, Excel worksheet 34.
See Excel file Case Study Derivatives Part 2, Excel worksheet 35.
See Excel file Case Study Derivatives Part 2, Excel worksheet 36.
See Excel file Case Study Derivatives Part 2, Excel worksheet 37.
See Excel file Case Study Derivatives Part 2, Excel worksheet 38.
See video "Straddle and Strangle."
See Excel file Case Study Derivatives Part 2, Excel worksheet 39.
See video "Straddle and Strangle."
See Excel file Case Study Derivatives Part 2, Excel worksheet 40.
See video "Strip, Strap and Gut."
See Excel file Case Study Derivatives Part 2, Excel worksheet 41.
See video "Strip, Strap and Gut."
See Excel file Case Study Derivatives Part 2, Excel worksheet 42.
See video "Strip, Strap and Gut."

Chapter 6
Overview of Option Strategies and Recommendations for Action

6.1 Assignment 37: Deriving the Optimal Strategy Based on the Basic Strategies as Well as Bullish and Bearish Strategies

Task

In order to be able to provide an assessment of which strategy is suitable for which market expectation of the investor, the considered value range of the Pharma Group share, from EUR 0.00 to EUR 140.00, is now divided into five classes.

Strong increase	140.00 EUR to 110.00EUR
Moderate increase	110.00 EUR to 85.00 EUR
Constant development	85.00 EUR to 55.00 EUR
Moderate depreciation	55.00 EUR to 30.00 EUR
Heavy depreciation	30.00 EUR to 0.00 EUR

Calculate the average of the results of all bullish and bearish option strategies *(basic strategies and advanced strategies but without the neutral strategies)* within each class, and mark the strategy with the highest and the lowest value for each class.

Content

In this assignment, you can now condense the results of all bullish and bearish option strategies calculated so far. This means that each possible expression of the Pharma Group share price between 0 euros and 140 euros is assigned a corresponding profit/loss in steps of five.

Supplementary Information The online version contains supplementary material available at https://doi.org/10.1007/978-3-031-85822-2_6.

D. Ernst, J. Häcker, *Derivatives*, Springer Texts in Business and Economics,
https://doi.org/10.1007/978-3-031-85822-2_6

Important Formulas

The formulas are used as shown in Figs. 4.3–4.6 and Assignment 8–18.

Display in Excel

The values in column D were determined by linking cell D14 to cell C20 in Worksheet 3 and then dragging it to D42. The same procedure was followed for columns F, H, J, L, N, P, R, T, V, X, Z, AB, AD, and AF.

The average values, shown in column E, were calculated as follows:

E14=> =MEAN(D14:D20)
E20=> =MEAN(D20:D25)
E25=> =MEAN(D25:D31)
E31=> =MEAN(D31:D36)
E36=> =MEAN(D36:D42)

This procedure was carried out for all bullish and bearish strategies, as well as the basic strategies, in the order presented in the previous assignments. Assignment 8 is considered to be mapped within the context of Assignment 7.

The following picture emerges, as shown in Fig. 6.1.

In the next step, the average values of the individual strategies are presented in an overview table as shown in Fig. 6.2.

In Fig. 6.2, the cells E5, E6, E7, E8, and E9 are linked to the cells E14, E20, E25, E31, and E36 in Fig. 5.18. The same procedure is followed for the other columns. The green-highlighted cells indicate which strategy is the best strategy in each case, depending on how the Pharma Group share price changes. Correspondingly, the red-highlighted fields reflect the worst strategy.

Fig. 6.1 Performance of the bullish and bearish strategies as well as the basic strategies depending on the change in the share price of the Pharma Group

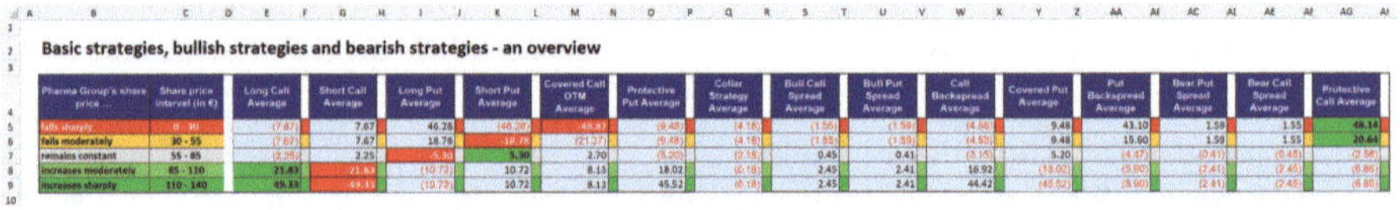

Basic strategies, bullish strategies and bearish strategies - an overview

Pharma Group's share price ...	Share price interval (in €)	Long Call Average	Short Call Average	Long Put Average	Short Put Average	Covered Call OTM Average	Protective Put Average	Collar Strategy Average	Bull Call Spread Average	Bull Put Spread Average	Call Backspread Average	Covered Put Average	Put Backspread Average	Bear Put Spread Average	Bear Call Spread Average	Protective Call Average
falls sharply	0 - 30	(7.67)	7.67	46.28	(46.28)	-45.87	(9.48)	(4.18)	(1.56)	(1.59)	(4.66)	9.48	43.10	1.59	1.55	48.14
falls moderately	30 - 55	(7.67)	7.67	18.76	-18.76	(21.37)	(9.48)	(4.18)	(1.55)	(1.59)	(4.55)	9.48	15.60	1.59	1.55	20.64
remains constant	55 - 85	(2.25)	2.25	-5.30	5.30	2.70	(5.20)	(2.18)	0.45	0.41	(3.15)	5.20	(4.47)	(0.41)	(0.45)	(2.58)
increases moderately	85 - 110	21.83	-21.83	(10.72)	10.72	8.13	18.02	(6.18)	2.45	2.41	16.92	(18.02)	(5.90)	(2.41)	(2.45)	(6.86)
increases sharply	110 - 140	49.33	-49.33	(10.72)	10.72	8.13	45.52	(6.18)	2.45	2.41	44.42	(45.52)	(5.90)	(2.41)	(2.45)	(6.80)

Fig. 6.2 Performance of the bullish and bearish strategies and the basic strategies aggregated into classes depending on the change in the share price of the Pharma Group

6.2 Assignment 38: Derivation of the Optimal Strategy Based on the Neutral Advanced Options Strategies

Task

In order to be able to provide an assessment of the *neutral* strategies, at which market expectation of the investor which strategy is suitable, the considered value range of the Pharma Group share from EUR 0.00 to EUR 140.00 is divided into the same five classes, analogous to the previous assignment.

Calculate the average of the results of all bullish and bearish *neutral* option strategies within each class and mark the strategy with the highest and lowest value for each class. Base your analysis on the 18 neutral strategies shown above (8 neutral-bearish and 10 neutral-bullish).

Content

The procedure in Assignment is analogous to the procedure in Assignment 37.

Important Formulas

The formulas are used as presented in the previous assignments.

Display in Excel

Analogous to Assignment 37, the following picture emerges, as shown in Fig. 6.3.

In the next step, the average values of the individual strategies are presented in an overview table as shown in Fig. 6.4.

Again, the green-highlighted boxes per class show the best strategy, and the red-highlighted boxes show the worst strategy.

6.3 Assignment 39: Final Derivation of the Optimal Strategy Based on the Basic Strategies and the Advanced Options Strategies

Task

In order to be able to provide an assessment for all strategies, at which market expectation of the investor each strategy is suitable, the considered value range of

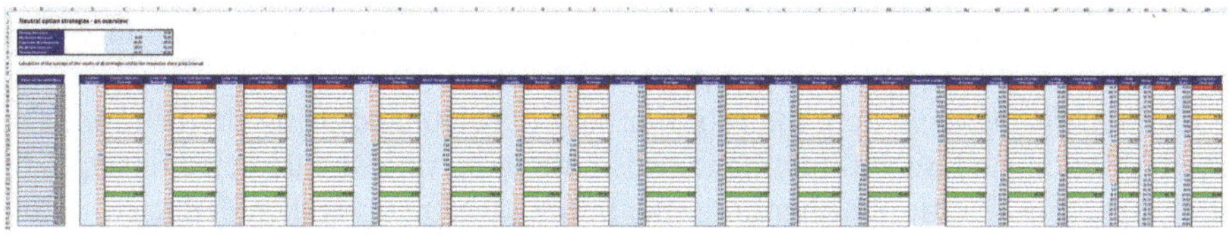

Fig. 6.3 Performance of neutral strategies aggregated into classes according to changes in the Pharma Group share price

Neutral option strategies - an overview

Pharma Group's share price ...	Share price interval (in €)	Condor Options Average	Long Call Butterfly Average	Long Put Butterfly Average	Long Call Ladder Average	Long Put Ladder Average	Short Strangle Average	Short Straddle Average
falls sharply	0 - 30	(0.22)	(0.07)	(0.07)	5.32	-43.93	(38.56)	(38.65)
falls moderately	30 - 55	(0.22)	(0.07)	(0.07)	5.32	-16.43	(11.06)	(11.15)
remains constant	55 - 85	0.07	0.21	0.21	3.03	4.78	7.58	7.78
increases moderately	85 - 110	(0.22)	(0.07)	(0.07)	-18.18	7.07	(11.06)	(11.15)
increases sharply	110 - 140	(0.22)	(0.07)	(0.07)	-45.68	7.07	(38.56)	(38.65)

Pharma Group's share price ...	Short Guts Average	Short Condor Strategy Average	Short Call Butterfly Average	Short Put Butterfly Average	Short Call Ladder Average	Short Put Ladder Average	Long Strangle Average
falls sharply	(38.60)	0.22	0.07	0.07	(5.32)	43.93	38.56
falls moderately	(11.10)	0.22	0.07	0.07	(5.32)	16.43	11.06
remains constant	7.54	(0.07)	(0.21)	(0.21)	(3.03)	(4.78)	(7.58)
increases moderately	(11.10)	0.22	0.07	0.07	18.18	(7.07)	11.06
increases sharply	(38.60)	0.22	0.07	0.07	45.68	(7.07)	38.56

Pharma Group's share price ...	Long Straddle Average	Strip Average	Strap Average	Long Guts Average
falls sharply	38.65	84.17	31.79	38.60
falls moderately	11.15	29.17	4.29	11.10
remains constant	(7.78)	-12.97	(10.35)	(7.54)
increases moderately	11.15	1.67	31.79	11.10
increases sharply	38.65	29.17	86.79	38.60

Fig. 6.4 Performance of neutral strategies aggregated into classes according to changes in the Pharma Group share price

the Pharma Group share from EUR 0.00 to EUR 140.00 is divided into the same five classes, analogous to the previous assignment.

Calculate the average of the results of all option strategies within the respective class and mark the strategy with the highest and the lowest value for the respective class. In contrast to the two previous assignments, you now take into account the different amounts of premium paid or received. So far, the advantageousness of the strategies has only been shown in absolute terms. Now, it is about the relative advantageousness of all option strategies. Assume that you have an initial budget of €100.

Content

Step 1: In the first step, present the results of Figs. 5.19 and 5.21 together in a summary table. You obtain the respective results by connecting the corresponding cells.

Step 2: Calculation of the profit/loss, including the IP ratio. It is assumed that the investor has €100 available for investment (I = investment). This initial investment is put in relation to the premium to be paid (P = premium) (IP ratio) and shows how often the investor can execute the respective option strategy. A negative number (initial investment) divided by a negative number (premium to be paid) then results in a positive IP ratio. If the effect of short positions predominates, the premium is positive, and the initial investment accordingly becomes an initial income. Accordingly, a positive number (initial income) divided by a positive number (premium received) then again gives a positive IP ratio.

Step 3: In the next step, the profit or loss is calculated using the IP ratio. The profit is calculated by multiplying the absolute profit or loss calculated in step 1 by the respective IP ratio. Furthermore, the initial investment is then deducted, or the initial income is added. Depending on the change in the Pharma Group share price, the best/worst option strategy is calculated and highlighted in green/red.

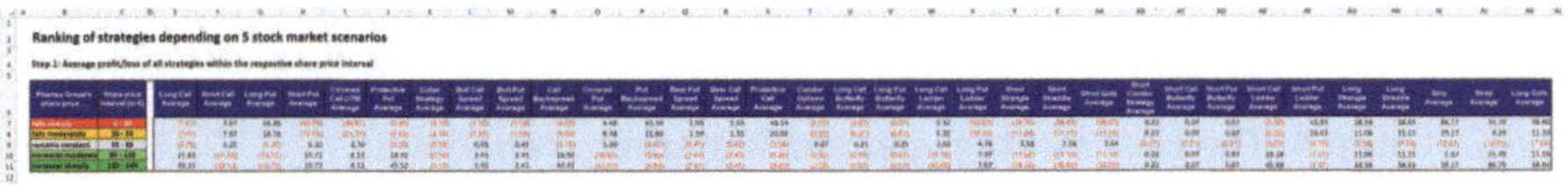

Fig. 6.5 Absolute performance of all strategies

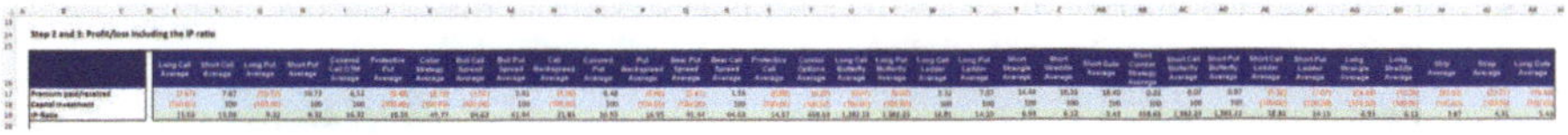

Fig. 6.6 Calculation of the IP ratio

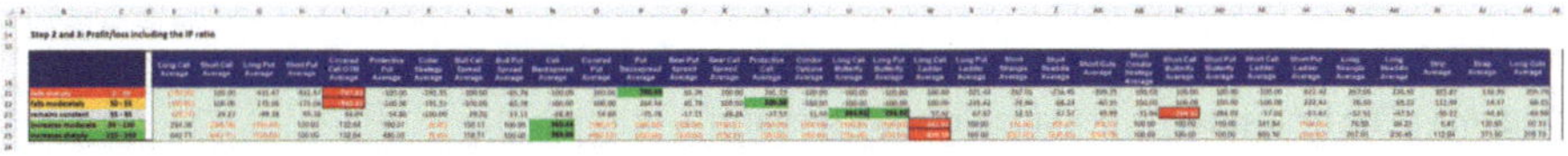

Fig. 6.7 Profit or loss including the IP ratio

Steps 4 and 5: The best or worst option strategy, depending on the change in the share price of the Pharma Group, is finally highlighted in a summary table.

Important Formulas

The formulas are used as presented in the previous assignments.

Step 1: For example, cell E7 results from linking to cell A5 in Fig. 5.19.

Step 2: The premium paid in cell E17 was linked to cell C50 in worksheet 3. The capital investment in cell E18 is obtained by linking to the initial investment in the input sheet and the corresponding sign of the premium. Thus:

E18 => =SIGN(E17)*Input!$C$81. The IP ratio is calculated by putting both quantities in proportion.

E19 => =ABS(E18/E17).

Step 3: The profit when the share price falls sharply, in cell E21, is calculated as follows:

E21 => =(E$19*E7)

The best option strategy is calculated in column AM, and the worst option strategy is calculated in column AO. For example:

AM21 => =MAX(E21:AK21)
AO21 => =MIN(E21:AK21)

Display in Excel

Step 1 (Fig. 6.5):

Step 2 (Fig. 6.6):

Step 3 (Fig. 6.7):

Step 4 (Fig. 6.8):

Step 5 (Fig. 6.9):

Step 4: The best strategy per stock market scenario

		Call Backspread Average	Put Backspread Average	Protective Call Average	Long Call/Put Butterfly
falls sharply	0 - 30		730.63		
falls moderately	30 - 55			300.59	
remains constant	55 - 85				294.92
increases moderate	85 - 110	369.43			
increases sharply	110 - 140	969.86			

Fig. 6.8 The best strategy per stock market scenario

Step 5: The worst strategy per stock market scenario

		Covered Call OTM Average	Short Call Butterfly Average	Long Call Ladder Average
falls sharply	0 - 30	-797.62		
falls moderately	30 - 55	-348.81		
remains constant	55 - 85		-294.92	
increases moderate	85 - 110			-341.94
increases sharply	110 - 140			-859.10

Fig. 6.9 The worst strategy per stock market scenario

6.4 Assignment 40: Application of Option Strategies to the Corona Crisis

Task

The following figure shows the development of the DAX before and during the Corona crisis. We go back to March 2020, when the Corona crisis had a clear impact on the stock market (see Fig. 6.10).

Put yourself in the real situation of an investor who wants to continue investing in options on the Pharma Group share. The respective situations are to be considered *ex ante*.

Fig. 6.10 The development of the DAX between September 2019 and September 2020

It is March 18, 2020, and the DAX stands at 8441 points. The spread of the coronavirus is progressing and the effects are very difficult to foresee. Which strategy do you choose? Give reasons for your answer. To answer this question, you can derive answers to the following 10 questions:

Question 1: Where does the Pharma Group share currently stand, and where does the DAX stand?

Question 2: What is the price/book ratio (P/B) of the Pharma Group share and the DAX? What does this imply?

Question 3: How do you assess the development of the Pharma Group's business segments?

Question 4: How do you assess the impact on the industries of Pharma Group's customers?

Question 5: What is the impact of the ECB's *monetary policy* measures on the crisis? Which measures have already been implemented?

Question 6: What *regulatory* measures have been taken?

Question 7: How do you assess the *general fiscal policy* measures (e.g., the suspension of insolvency, state participation in companies, decisions regarding short-time work)?

Recommendations for action depending on the current stock market situation

Initial situation	
Date	March 18, 2020
DAX level	8,441.52
Volatility	High

Fig. 6.11 Recommendations for action depending on the current stock market situation

Assessment of the development of the Pharmaceuticals Group's business areas

Business field	Entwicklung
Generic	↗
Consumer Health	↑
Oncology	↗
Nutrition and health	↑

Overall assessment	↑

→ : constant
↗ : moderate increase
↑ : strong increase
↘ : modeate decrease
↓ : strong decrease

Fig. 6.12 Assessment of the development of the Pharma Group's business areas

Question 8: How high is the volatility on the market to be estimated?

Question 9: Which option strategy do you choose?

Question 10: What is your cash position expected to be on March 18, 2020?

Content

Ad Question 1 (Fig. 6.11):

Ad Question 2:

The price/book ratio of the DAX was 0.9 at the low point in 2003 and 1.01 at the low point in 2009. In business valuation, it is true that, in the long term, the book value forms the lower limit of value. In Mergers & Acquisitions, it is said: "The seller does not sell below book." What is true for one company is also true for 40 companies. If the 40 price/book ratios are added up (market capitalization / (equity – minority interests)) and divided by 40, it can be seen that, at this point in time, it is again true that market capitalization is approximately equal to book value.

Since the beta of the Pharma Group is 1, the share price of the Pharma Group is analogous to the DAX. The Pharma Group share price, therefore, has enormous upside potential.

Ad Question 3 (Fig. 6.12):

Impact on the industries of the Pharma Group's customers?

Customer industries	Entwicklung
Wholesalers for pharmaceuticals	↗
Retail chains	→
Government-sponsored health care systems	↑
Providers of Managed Care	↗
Private healthcare systems	↗

Overall assessment	↗

Fig. 6.13 Impact on the industries of the Pharma Group's customers

Impact of the central bank's MONEY POLICY: What measures have already been implemented?

1. ECB interest rates unchanged at:*	
- Main refinancing rate: 0.00%	0.0%
- Deposit facility: -0.5%	-0.5%
- Marginal lending facility: 0.25%	0.25%
The ECB announces a Pandemic Emergency Purchase Program (PEPP) worth €750 billion.	750,000,000,000
3. FED: Reduction of the prime rate to 0.00%.	0.00%

** European Central Bank*

Fig. 6.14 Impact of the money policy

Impact of REGULATORY POLICIES: What measures have already been implemented?

- In particular, macro potential policy. The micro potential policy can be ignored here. The efficiency of monetary policy is increased by regulatory policy.

Countercyclical capital buffer - CCyB: BAFin lowered the domestic CCyB quota to 0%.**	0.0%

*** finanzwende.de*

Fig. 6.15 Effects of the regulation policy

Ad Question 4 (Fig. 6.13):
Ad Question 5 (Fig. 6.14):
Ad Question 6 (Fig. 6.15):
Ad Question 7 (Fig. 6.16):
Ad Question 8 (Fig. 6.17):
Ad Question 9:

Based on a DAX P/B multiple of about 1, it can be assumed that the DAX has bottomed out. The VDAX is very high, and a rebound could occur soon. Buying calls, therefore, seems advantageous—although a stock market rule says: "Never reach into a falling knife." To profit from the expected rebound, two calls can be

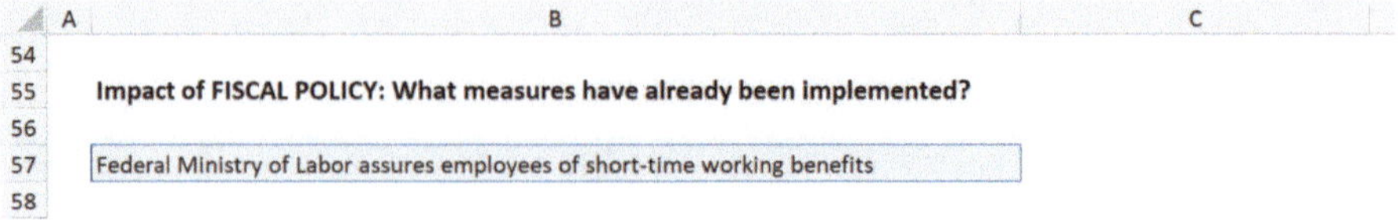

Fig. 6.16 Effects of the fiscal policy

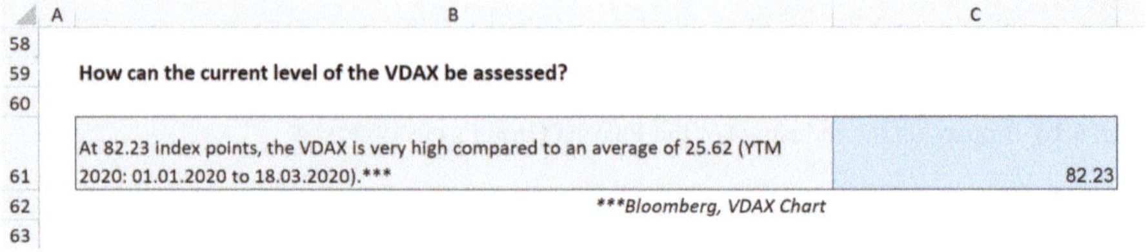

Fig. 6.17 Assessment of market volatility

bought. To hedge on the downside, a put can also be bought. The investor could be wrong, or the rebound could start later. Here, too, the ex-ante perspective is still the basis.

Ad Question 10:

The question now arises as to how many euros have become of the initial position of €100 on March 18, 2021. Within the option strategy Strap, two calls and one put are bought. Let us now leave the ex-ante perspective and take the ex-post perspective. The share price of the Pharma Group has risen to €122.52. This results in an intrinsic value per call of €52.52. Since two calls were bought, this position has an intrinsic value of €105.04. The premium paid of €23.21 is to be deducted from this, resulting in a profit per strap of €81.83. Due to the significant increase in the share price, the put expires. With the initial investment of €100, 4.31 straps can be realized arithmetically. *This results in an expected cash position of €352.53 on March 18, 2021.*

If only Pharma shares had been bought from the €100, the cash position would have been only €175.03 (the share increased by 75.03% – ((122.52/70) – 1)). If only calls on Pharma Group shares had been bought from the €100, the cash position would have been €665.06 (€6.86 premium; €52.52 intrinsic value; number of calls bought: 14.57; profit per call: €45.66; total profit = €45.66 * 14.57). In hindsight, however, one is always smarter. That is the subtle difference between ex-ante and ex-post.

Important Formulas

C72 => =Input!C67
C115 =>=Input!C68
C116 =>=-C76*C75-C79*C78
C117 =>=ABS(C115/C116)
C118 =>=C117*C108

Pharma Group	
Share price (as of March 18, 2020)	70.00
Share price (as of March 18, 2021)	122.52

Assumptions	
# Long Calls	2.00
Long Call Price	6.86
Strike price of Long Call	70.00
# Long Puts	1.00
Long Put Price	9.48
Strike price of Long Put	70.00

Price of the underlying	Profit/loss
0.00	46.79
5.00	41.79
10.00	36.79
15.00	31.79
20.00	26.79
25.00	21.79
30.00	16.79
35.00	11.79
40.00	6.79
45.00	1.79
50.00	(3.21)
55.00	(8.21)
60.00	(13.21)
65.00	(18.21)
70.00	(23.21)
75.00	(13.21)
80.00	(3.21)
85.00	6.79
90.00	16.79
95.00	26.79
100.00	36.79
105.00	46.79
110.00	56.79
115.00	66.79
120.00	76.79
122.52	81.83
130.00	96.79
135.00	106.79
140.00	116.79

Cash and profit	
Cash position (as of March 18, 2020)	100.00
Investment using Strap (as of March 18, 2020)	(23.21)
# Strap investments	4.31
Cash position (as of March 18, 2021)	352.53

Fig. 6.18 Calculation of the cash position on March 18, 2020

Display in Excel
Ad Question 10 (Fig. 6.18):

Further Reading

See Excel file Case Study Derivatives Part 2, Excel worksheets 43 and 44.
See video "Aggregation."
See Excel file Case Study Derivatives Part 2, Excel worksheets 45 and 46.
See video "Aggregation."
See Excel file Case Study Derivatives Part 2, Excel worksheet 47-51.
See video "Aggregation."
See Excel file Case Study Derivatives Part 2, Excel worksheet 52-60.
See "Applying Option Strategies" video.

MIX
Papier aus verantwortungsvollen Quellen
Paper from responsible sources
FSC® C105338

If you have any concerns about our products,
you can contact us on
ProductSafety@springernature.com

In case Publisher is established outside the EU,
the EU authorized representative is:
Springer Nature Customer Service Center GmbH
Europaplatz 3, 69115 Heidelberg, Germany

Printed by Libri Plureos GmbH
in Hamburg, Germany